FOOD SCIENCE AND TECHNOLOGY

VISCOSITY OF FRUIT JUICES: EXPERIMENTAL AND MODELING

FOOD SCIENCE AND TECHNOLOGY

Additional books in this series can be found on Nova's website under the Series tab.

Additional E-books in this series can be found on Nova's website under the E-books tab.

FOOD SCIENCE AND TECHNOLOGY

VISCOSITY OF FRUIT JUICES: EXPERIMENTAL AND MODELING

I. M. ABDULAGATOV
M. A. MAGERRAMOV
A. I. ABDULAGATOV
AND
N. D. AZIZOV

Nova Science Publishers, Inc.
New York

For permission to use material from this book please contact us:
Telephone 631-231-7269; Fax 631-231-8175
Web Site: http://www.novapublishers.com

NOTICE TO THE READER

LIBRARY OF CONGRESS CATALOGING-IN-PUBLICATION DATA
Viscosity of fruit juices : experimental and modeling / I. M. Abdulagatov
... [et al.].
p. cm.
Includes bibliographical references and index.
ISBN 978-1-61209-525-7 (softcover : alk. paper)
1. Fruit juices--Viscosity--Measurement. 2. Fruit juices--Thermal properties. 3. Arrhenius equation. I. Abdulagatov, I. M.
TP562.V57 2011
536'.413--dc22

2011003984

Published by Nova Science Publishers, Inc. † New York

CONTENTS

PREFACE

Modeling, optimization and automation of food processes is difficult due to the complexity of the raw materials and products involved, which affect thermophysical properties such as density, heat capacity, viscosity, thermal conductivity and thermal diffusivity. The thermophysical properties of fruit juices exhibit substantial changes with temperature, pressure and concentration during processing. This new book presents current research on the effect these factors on fruit juices.

ABSTRACT

Viscosity of four fruit (pomegranate, pear, tangerine and lemon) juices have been measured at temperatures from 292 to 403 K and at pressures from 0.1 to 10 MPa for the concentrations from 11 to 45 °Brix. The best and complete comprehensive compilations all of the available viscosity data for liquid foods at the present time are provided. The effect of temperature, pressure, and concentration on the viscosity fruit juices were studied. The measured values of the viscosity were used to calculate the temperature, $(\partial \ln \eta / \partial T)$, and concentration, $(\partial \ln \eta / \partial x)_T$, coefficients for the fruit juices. Various theoretical models (polynomials, power, exponential, logarithmic, and their various combinations) for the viscosity of fruit juices were stringently tested with new accurate measurements on pomegranate, pear, tangerine and lemon juices. The applicability, accuracy and predictive capability of the various models were studied. New models were developed to accurately represent the combined effect of temperature and concentration on the viscosity of the fruit juices. Models which represent the viscosity of fruit juices relative to pure water viscosity was considered. The empirical extension of the various theoretically based models for the viscosity of aqueous solutions to the fruit juices is considered. It was found that theoretically based models originally developed for the viscosity of aqueous system can be adopted with success for fruit juices. The average absolute deviation (AAD) between measured and calculated values from these models for the viscosity of fruit juices was within 0.8 % to 2 %.

The values of the Arrhenius equation parameters (flow activation energy, E_a, and η_0) were calculated for the measured viscosities of pomegranate, pear, tangerine, and lemon juices as a function of concentration. The new viscosity models for the fruit juices are recommended for the future scientific and engineering used.

Keywords: *Arrhenius equation; pomegranate, pear, tangerine, lemon juices; viscometer; viscosity*

Chapter 1

INTRODUCTION

Modeling, optimization and automation of food processes is difficult due to complexity of the raw materials and product involved, which affect thermophysical properties such as density, heat capacity, viscosity, thermal conductivity, and thermal diffusivity. The thermophysical properties of fruit juices exhibit substantial changes with temperature, pressure, and concentration during processing (storage, transport, marketing and consumption, chilled, change temperature, tank farm change concentration, evaporator change concentration), see for example, Moressi and Spinosi [1] and Crandall et al. [2]. For this reason the thermophysical properties (density, heat capacity, viscosity, thermal conductivity, and thermal diffusivity) should be studied as a function of temperature, pressure, and concentration. Last 20 years the high-temperature and high-pressure technologies (high-temperature and high-pressure treatment in food preservation and pasteurization) was expanded to food industry. The preservation temperature is about 60 to 90 °C. Many fluid foods are subjected to high temperatures (above 90 °C) and high pressure (up to 350 MPa) during pasteurization and thermal processing [3-9] (high pressure food processing technologies and other industrial operations) . High pressure presents unique advantages over conventional thermal treatments including, application at low temperature, which improves the retention of food quality. Almost all previous thermophysical properties measurements for liquid foods were performed at atmospheric pressure, although high pressure food processing technologies requires accurate knowledge of thermophysical properties at moderate (3-10 MPa) and high pressures (up to 350 MPa). Therfeore, viscosity (flow properties) and other thermophysical properties of liquid foods at high temperatures and high pressures are need in these processing conditions. Very few measurements are available in the literatures at high temperatures. A literature survey revealed that there are no viscosity and other thermophysical data for liquid foods under prerssure, except our recent density, viscosity, and thermal conductivity measurements for fruit juices (Magerramov [10] and Magerramov et al. [11-15]). Little is known about the effect of temperature, pressure, and concentration on the thermophysical properties of liquid foods.

Knowledge of the viscosity is of primary importance to the fruit juice industry. For example, viscosity is an important property of juices in the prediction of heat – and mass-transfer coefficients and in the design of heat- and mass-transfer equipment for fruit juice industry. The accurate viscosity data and their variation with operating conditions (over wide temperature and concentration regions) are need also for a various research and engineering applications in any branches of the food industry, such as developing food processes and processing equipment (detailed design and evaluate food processing equipment such as pumps, heat exchangers, evaporators, equipment sizing, filters and mixers, engineering calculations on evaporation rates, pumping and pipe requirements), control of products, quality evaluation and an understanding of the structure of food and raw agricultural materials. Viscosity changes are determinant factors in operations such as concentration by evaporation and reverse osmosis, pumping, homogenization and blending.

Variations in the viscosity of fruit juices affect the energy usage in a fruit processing plant (Crandall et al. [2]). The type of evaporator, direction of feed, and heat transfer rate are all affected by viscosity. High shear rates are utilized in modern evaporators to reduce the viscosity, increase the heat-transfer rate, and thus safe energy. If the viscosity of the concentrate exceeds a threshold value then the output products concentration must be reduced or the concentrate will "burn on" the inside of the evaporator (Crandall et al. [2]). This would cause a loss of energy and product. Viscosity can become an important factor during the concentration juices, especially in the production of high density concentrates, due to the inefficiency of the operation when the product becomes highly viscous. Viscosity of fruit juices changes with content of soluble and suspended solids. Pectin and sugar concentration are the main factors in changes of viscosity.

For engineering utility, reliable methods for estimation of the viscosity and other thermophysical properties of fruit juices over wide ranges of concentration, temperature, and pressure would be extremely valuable. To understand and control those processes which used liquid foods, it is necessary to accurately know their thermodynamic (density, heat capacity, vapor pressure, activity coefficient) and transport (viscosity, thermal conductivity, and diffusivity) properties. However, the lack of reliable data for fruit juices over wide temperature, pressure, and concentration ranges makes it necessary to estimate the missing properties by empirical and semi-empirical prediction methods. Therefore, a new accurate experimental data for viscosity of fruit juices at high temperatures and high pressures needed also to improve and extend the range of validity of available estimation and correlation methods which will be capable of reproducing the experimental viscosity data within their uncertainty and develop new more reliable prediction techniques. Unfortunately, the thermophysical properties of liquid food products cannot be accurately predicted theoretically, due to their complicated physical and chemical structure. Therefore, the thermophysical properties measurements of liquid foods are also research interest. The available theoretical models for liquids cannot describe complex real system as they met in practice. Better prediction models can be developed based on reliable experimental information on thermophysical properties liquid foods. Thus, there is great practical and theoretical interest in the study of the effect of temperature, pressure, and concentration on the thermophysical properties fruit juices at equipment operating conditions. The main objectives of this study are: (a) review of the often used techniques for the viscosity measurements for liquids; (b) present for the readers the comprehensive review all of the available experimental viscosity data for liquid foods (viscosity data compilation); (c) experimental and theoretical study of the effect of temperature, pressure, and concentration on the viscosity behavior of fruit juices; (d) examine the accuracy and predictive capability various empirical, semi empirical, and predictive models for the viscosity of liquid foods; (e) develop accurate models for the combined effect of the temperature and concentration on the viscosity of fruit juices; (f) provide for the readers the recommendations of the best viscosity models for fruit juices for the future scientific and engineering used. Available experimental thermophysical properties (density, heat capacity, viscosity, thermal conductivity, and thermal diffusivity) data for liquid foods have been reviewed previously by various authors [16-32]. We also provided a new accurate experimental viscosity data for four fruit (tangerine, lemon, pomegranate, and pear) juices at high temperatures (up to 120 °C), at pressures up to 10 MPa and at concentrations up to 45 °Brix using a capillary-flow techniques, which have been previously used for the accurate measurements on other fluids [33-37] at high temperatures and high pressures. The present results considerably expand the temperature, pressure, and concentration ranges in which viscosity data for these fruit juices are available.

Table 1. Physico-chemical characteristics of fruit juices.

Pear juice		Pomegranate juice	
Soluble solids	15.2 °Brix	Soluble solids	11°Brix
Pectin	0.25 %	Citric acid	1 %
Total sugar	8.70 %	Limon acid	1 %
Glucose	1.43 %	Glucose	6 %
Fructose	6.91 %	Sucrose	-
Sucrose	0.36 %	Acidity	-
Amino acid nitrogen	0.141 %	Fructose	7 %
Tannic acid	0.0171	Potassium*	1100
Cellulose	0.90	Chlorides*	500
pH	4.15	Magnesium*	-
Potassium	48 mg l^{-1}	Phosphates*	300
Calcium	12 mg l^{-1}	Other minerals	5
Magnesium	3 mg l^{-1}	-	-
Phosphate	13 mg l^{-1}	-	-
Ash	0.3	-	-
Lemon juice		Tangerine juice	
Soluble solids	17 °Brix	Soluble solids	15 °Brix
Total sugar	3.0 %	Total sugar	5.4 %
Sucrose	-	Glucose	1.1 %
Glucose	-	Fructose	3.8 %
Fructose	2.9 %	Sucrose	0.4 %
Pectin	-	Acidity	-
Potassium*	30	Potassium*	33 mg
Calcium*	6	Calcium*	6 mg
Magnesium*	5	Magnesium*	6 mg
pH	2.4	pH	3.5

* mg in 100 g juice.

1. Material Descriptions

Experimental samples 11.0, 15.2, 15.0, and 17.0 °Brix of pomegranate, pear, tangerine, and peach juices used in this study were obtained from fresh full-ripe fruits from a plant in Baku, Azerbaijan. The natural juices were obtained by squeezing the full-ripe fruits with a laboratory screw press, eliminating the suspended solids by filtering and clarifying. Concentrated juices with various soluble solids contents were obtained from the original concentrate using a rotary glass vacuum evaporator (SPT-200, Zeamil-Horyzont, Poland) at temperature below 60 °C. The evaporation chamber was rotated at a constant rotational speed in water bath at 40°C. The soluble solids content as °Brix was measured using a universal laboratory refractometer

(RLU-1, Russia) at room temperature (20 °C). In order to adjust the concentration of the juice, the concentrated juice was diluted with distilled water. The samples were stored in glass vessel at 2-4 °C (8 hours) until used for the density measurements. Microelements (potassium, calcium, magnesium, and phosphates) were determined using an atomic absorption spectrophotometer (C-115-M1, Russia). The glucose and fructose contents were determined by the method of Bertrand. The total sugar was calculated by summation of individual sugars. The pH was measured using a digital pH-meter (Kent EIL 7020, UK) at 20 °C. Total acidity was determined by potentiometric titration with NaOH 0.1 N until pH 8, monitored with pH-meter. Physical and chemical characteristics of pomegranate, pear, tangerine, and lemon juices are given in Table 1.

Chapter 2

VISCOSITY MEASUREMENTS

The viscosity of a liquid is known to determine the relationship between the gradients of pressure and corresponding flow. However, this flow enters hydrodynamic equation which contains nonlinear terms. The response of the system to the presence of finite pressure gradient is so intricate that it is difficult to take this gradient into account in experiment. For example, the pressure gradient induces secondary flows and energy dissipation in addition to laminar flow. This difficult is challenge researcher to develop special techniques for measuring of viscosity which would diminish the effect to a minimum.

For fluids that obey Newton's Law, the dynamic viscosity of a fluid is a measure of its tendency to dissipate energy when it is disturbed from equilibrium by a velocity field v. The dynamic viscosity, η, is thus defined by the relationship

$$\sigma_{ij} = -P\delta_{ij} + \eta\left(\frac{\partial v_i}{\partial x_j} + \frac{\partial v_j}{\partial x_i} - \frac{2}{3}\delta_{ij}\frac{\partial v_k}{\partial x_k}\right). \quad (1)$$

In this equation, σ_{ij} are the instantaneous stresses, P, the pressure and δ_{ij} is the Kronecker symbol. The viscosity depends on the thermodynamic state, (T, P, x) or (T, ρ, x), of the fluid and fluid mixture (where T is the temperature, ρ the density, and x the concentration).

2.1. Experimental Techniques

The definition of the dynamic viscosity given by Eq. (1) presumes the ability to measure local shear stresses. Since however, it is impossible to measure local shear stresses, and viscometers undoubtedly do not have a constant shear rate, methods of measurement of the viscosity must be based on the determination of some integral effect of the stresses amenable to precise measurement in a known flow field. Inevitably, the imposition of a shear field generates small pressure differences, and dissipation causes local temperature gradients, both of which change slightly the reference thermodynamic state to which a measurement is assigned, from the initial, unperturbed, equilibrium state. The reference state for the measurement will be obtained by averaging so it is important that the system is disturbed as little as possible from equilibrium during measurement.

Techniques frequently employed to measure the viscosity of liquids are: 1. capillary flow; 2. oscillating discs; 3. oscillating cups; 4. falling body; 5. rotating cylinders; 6. torsional crystal oscillator; and 7. vibrating wire. In Table 2 all available measurements, to our knowledge, of the viscosity of fruit juices are presented. It can be seen that almost all data were derived by a capillary-flow technique and coaxial cylinder, while only very few datasets were obtained by the falling-body technique and rotational viscometer.

Table 2. Summary of experimental viscosity data for liquid foods.

Liquid food	Authors	Concentration	Temperature (°C)	Method
Cranberry	Peleg and Noble [185]	12 °Brix	5-25	CC
Passion fruit	Vitali et al. [186]	15.6-33.4 °Brix	20-50	CC
Passion fruit, peanapple, banana, guava, papaya	Garciá et al. [187]	12.6-26.5 °Brix	22-40	CV
Apple	Geller et al. [214]	12-60 °Brix	20	CC
Apple, grape,	Saravacos [160]	19-70 °Brix	27	CV
Apple, grape, orange,	Saravacos [38]	10.5-75° Brix	20-70	CC
Apple, grape	Rao [112]	15-75° Brix	-5-41	RV
Orange	Barros [109]	54.99-68.83 °Brix	30	-
Orange	Steverson [163]	-	30	-
Orange	Moresi and Spinosi [162]	9.6-65 °Brix	25-50	FSV

Liquid food	Authors	Concentration	Temperature (°C)	Method
Orange, grapefruit	Ezell [167]	23-67	30	RV
Orange	Vitali and Rao [113]	65 °Brix	-19-30	CC
Orange	Hernandez et al. [169]	12.2-66.5 ° Brix	5-75	CV
Orange	Mizrahi and Berk [164]	11-60 °Brix	30-70	CV
Orange	Mizrahi and Berk [165]	60-65 °Brix	30-80	CC
Orange	Dubois and Murdock [170]	42, 58.5 °Brix	-1.1-23.3	-
Orange, peanapple	Varshney and Kumbhar [168]	7-36 °Brix	30-45	CC
Orange	Crandall et al. [2]	12.8-65 °Brix	-10-25	CC
Orange	Berk [171]	60 °Brix	-	CV
Orange	Ibarz et al. [101]	30.7-63.5 °Brix	5-70	CC
Orange (concentrate)	Mizrahi and Firstenberg [166]	1.5-11°Brix	30	CC
Orange	Rouse et al. [39]	57.3-65.1 °Brix	25.6	CV
Orange	Rouse et al. [172]	45 °Brix	25.6	CV
Orange	Ingram [173]	63-69.5°Brix	15-45	-
Grape	Bayindirli [95]	19-35 °Brix	20-80	CV
Grape	Moresi and Spinosi [174]	0-73.1 °Brix	20-50	RV
Grape musts	Lopez et al. [188]	990-1100 $kg \cdot m^{-3}$	14-22	RV, CC
Raspberry	Ibarz and Pagán [104]	15-30°Brix	5-60	CC
Peach	Ibarz et al. [105]	40-69°Brix	5-60	CC
Loquat	Ibarz et al. [102]	40-72°Brix	5-65	CC
Sloe	Ibarz et al. [103]	25-58.5°Brix	5-65	CC
Blackcurrant	Ibarz et al. [106]	35-64.5 °Brix	5-60	CC
Sea buckthorn	Oomah et al. [189]	48°Brix	25-80	SR
Pear, apple, grape, lime, plum, tomato, pineapple, watermelon, babaco, orange, lemon, tangerine	Alvarado and Romero [92]	8.4-70 °Brix	10-80	CV
Pear	Ibarz et al. [108]	40-71 °Brix	5-60	CC
Apple	Gochiyaev [159]	15-57 °Brix	23-65	CV

Table 2. (Continued).

Liquid food	Authors	Concentration	Temperature (°C)	Method
Apple	Constenla et al. [98]	12-68.5 °Brix	20-80	CC
Apple	Bayindirli [93]	14-39 °Brix	20-80	CV
Apple	Rao et al. [110]	51-70 °Brix	23-43	CC
Apple, pear	Ibarz et al. [107]	70, 71 °Brix	5-60	CC
Apple	Schwartz and Costell [161]	-	-	-
Celery	Lau et al. [190]	0-30 %	-9.1-25	ROV
Tomato	Fito et al. [176]	5-28 °Brix	15-70	RV
Tomato	Foda and McCollum [177]	-	35-95	RV
Tomato	Molwane and Gunjal [180]	6-18 °Brix	30-80	CC
Tomato	Krapivin and Fedotkin [181]	8-35 °Brix	25-70	CV
Tomato	Harper and El-Sahrigi [179]	5.8 - °Brix	32-82	CC
Tomato	Luh et al. [182]	6.67	20	RV
Tomato	Saravacos et al. [178]	9-32 °Brix	25	CV
Tomato	Alviar et al. [217]	11.78-35.74 °Brix	23	SR
Tamarind	Manohar et al. [191]	7-62 °Brix	25-70	CC
Liquorice extract	Maskan [192]	3-50 °Brix	10-60	RV
Malus floribunda	Cepeda and Villaran [97]	17-70 °Brix	25	RV
"Baneshan" and "Neelum" Mangoes	Gunjal and Waghmare [193]	20 °Brix	40-80	SV
Watermelon	Sogi [194]	23.59-42 °Brix	5-50	RV
Gilaboru	Altan et al. [158]	35-59.7 °Brix	5-60	SR
Strawberry	Juszczak and Fortuna [99]	50-67.1 °Brix	10-60	CC
Strawberry	Simpson [195]		22.8	CV
Cherry	Juszczak and Fortuna [100]	50-63.8 °Brix	10-60	CC
Cherry	Giner et al. [183]	22-74 °Brix	5-70	CC
Sour cherry	Bayindirli and Özsan [94]	13.8-26.1 °Brix	20-70	CV
Banana	Khalil et al. [175]	20-79.7 °Brix	30-70	CC

Liquid food	Authors	Concentration	Temperature (°C)	Method
Pomegranate	Kaya and Sözer [114]	45.3-71 °Brix	5-60	SR
Pomegranate	Altan and Maskan [184]	17.5-75 °Brix	10-55	SR
Pomegranate, pear	Magerramov et al. [14]	11-40 °Brix	19-130	CV
Pomegranate	Bayindirli et al. [96]	0.5, 1.0, 1.5, 2.0 g/L gelatin		CV
Tangerine, lemon	Magerramov et al. [15]	15-40 °Brix 17-45 °Brix	30-119 30-119	CV
Guava	Vitali and Rao [196]	9.8-16 °Brix	25-60	CV
Guava	Brekke and Myers [197]	7.2-40.8°Brix	24	CV
Corn sirup	Erickson et al. [128]	20-85 D.S.	15.6-82.2	FSV
Corn sirup, silicone oil	McCarthy and Seymour [218]	-	-	CM
Mango	Manohar et al. [198]	16-30 °Brix	30-70	CC
Mango concentrates	Rao et al. [111]	11-26 °Brix	25-62	CC
Mango	Singh and Eipeson [199]	15-66 °Brix	15-85	RV
Mulberry	Kaya [115]	46-72 °Brix	10-60	SR
Apricot	Watson [200]	22 °Brix	4.4-25	CC
Pear purees	Harper and Lebermann [201]	18-46 °Brix	32-82	CC
Guar gum and locust bean gum solution	Elfak et al. [202]	20 wt % Glucose, 40 wt % water; 20 wt % sucrose, 40 wt % water;	25	RV
Silicon and mineral oils	Aution and Houska [216]	-	20, 30	RV
Cassava starch pastes	Odigboh and Mohsenin [213]	2-8 wt %	50-95	RV

* CV-capillary viscometer; FSV-falling sphere viscometer; CC-coaxial cylinders; SR-stress rheometer; RV-rotational viscometer; ROV-routine viscometer; SV – synchrolectric viscometer; CM-consistometer.

In this section a brief analysis of these methods is presented. The theoretical basis of the methods, and the working equations employed are presented, together with a brief description of the experimental apparatus and the measurements procedure of each technique. For a more thorough discussion of the various techniques employed for the viscosity measurements of liquids and liquid foods, the reader is referred to the works [18,19,25,26,29,41].

2.1.1. Capillary-Flow Technique

As seen in Table 2, capillary viscometers are the most extensively used instruments for the measurement of viscosity of fruit juice. They have the advantage of simplicity of construction and operation.

2.1.1.1. Theoretical

The principle of the capillary viscometer is based on the Hagen-Poiseuille equation of fluid dynamics (first formulated by Hagenbach [42] in 1860) and its alteration for practical viscometry by Barr [43] in order to include the so-called kinetic-energy correction and the end correction. The resulting equation expresses the viscosity of a fluid flowing through a thin capillary, in terms of the capillary radius, *R*, the pressure drop along the tube, ΔP, the volumetric flow rate, *Q*, and the length of the tube, *L*, as

$$\eta = \frac{\pi R^4 \Delta P}{8Q(L+nR)} - \frac{m\rho Q}{8\pi(L+nR)} \tag{2}$$

where *n* is the end-correction factor and *m* is the kinetic-energy correction factor with $L >> R$. The above equation is derived on the assumptions that, (a) the capillary is straight with a uniform circular cross section; (b) the fluid is incompressible and Newtonian and (c) the flow is laminar and there is no slip at the capillary wall - to avoid turbulence (Reynolds number, Re, over 2000), viscometers are designed to operate where the Reynolds number is well bellow 300. The correction factors *n* and *m* reflect the fact that in a practical viscometer two chambers must be placed at either end of the capillary in order to measure the pressure drop. Thus, for example, the parabolic velocity distribution characteristic of most of the flow can only be realized some distance downstream from the inlet of the capillary.

In the range of Reynolds numbers between 0.5 and 100, the theoretical values of the correction factors *m* and *n* are represented as $m = m_0 + 8n/Re$ with $m_0 = 1.17 \pm 0.03$ and $n = 0.69 \pm 0.04$ [44]. The values of *m* and *n* can also be determined experimentally. Other corrections - capillary shape and slip, non-uniform cross-sections, elliptical cross-section, coiled capillaries, wall roughness, fluid properties effects, compressibility, non-Newtonian [45], and surface tension correction can be found in the respective literature [41,45].

2.1.1.2. Experimental

Figure 1 shows a schematic diagram of a typical capillary viscometer (Abdulagatov and Azizov [46]). A very similar capillary viscometer, differing only in the actual capillary dimensions, was developed by other authors (see [47-51]). The working capillary-1 with ID of 0.3 mm and length of 216 mm is made from stainless steel. The capillary-1 is soldered to the extension tube-6. The fluid under study flows to a cold zone through the extension tube-6. Capillary-1 with extension tube-6 located in the high temperature and high pressure autoclave-4. The extension tube-6 is connected to a movable cylinder-9, which in turn is connected to the fixed cylinder-11 by means of the flexible tube-10. Both cylinders (9 and 11) are supplied with identical expanded bottles, employed to stabilize the fluid efflux through the capillary. The input and output sections of the capillary have conical extensions. All parts of the experimental installation in contact with the sample are made out of stainless steel.

The capillary tube-1 is filled with the fluid and when the movable cylinder is moved vertically at constant speed, the fluid flows through the capillary. To create and also measure accurately the pressure, the autoclave is connected to a dead-weight pressure gauge by means of separating vessel-13. Mercury is used as the separating liquid.

2.1.1.3. Working Equation and Uncertainty

In the capillary viscometer shown in Figure 1, if we denote by t the time required for a volume of fluid V to flow through the capillary, and take into consideration the variation of the geometrical size of the capillary, and that of mercury and sample densities, then Eq. (2) becomes

$$\eta = \frac{\pi R^4 \Delta P t}{8(L+nR)V}(1+\alpha\Delta T)^3 - \frac{m\rho V}{8\pi(L+nR)t}, \tag{3}$$

where, α is the linear expansion coefficient of the capillary material, ΔT the temperature difference between experimental temperature and room temperature.

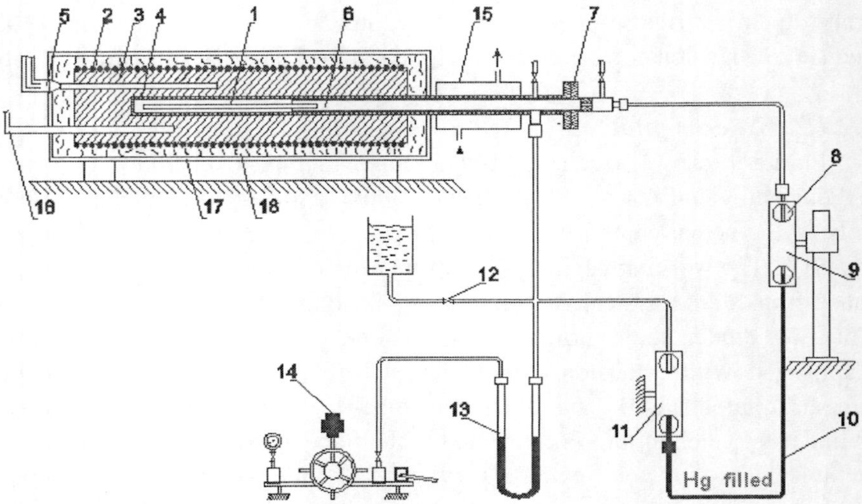

Figure 1. Schematic diagram of the experimental apparatus employed for the measurement of viscosity of aqueous solutions at high temperatures and high pressures by the capillary method (Abdulagatov and Azizov [46]). Working capillary-1; two electrical heaters-2; solid (substantial) red copper block-3; high pressure autoclave-4; thermocouple-5; extension tube-6; flange-7; viewing windows-8; movable cylinder-9; flexible connecting tube-10; unmovable (fixed) cylinder-11; valve-12; separating vessel-13; dead-weight pressure gauge (MP-600)-14; PRT-15.

The pressure drop, ΔP, is determined from the relation

$$\Delta P = g\,\Delta H(\rho_{\mathrm{Hg}} - \rho) \tag{4}$$

where, ΔH, is the average mercury level drop in the movable and fixed cylinders at the beginning and ending of the measurements, ρ_{Hg}, is the density of mercury at room temperature and experimental pressure, and ρ is the density of the fluid under study. Mercury level in the cylinders was measured with a cathetometer. Equation (3) can be simplified to

$$\eta = c_1 t - \frac{c_2}{t}. \tag{5}$$

where constants c_1 and c_2 can be obtained from Eq. (3).

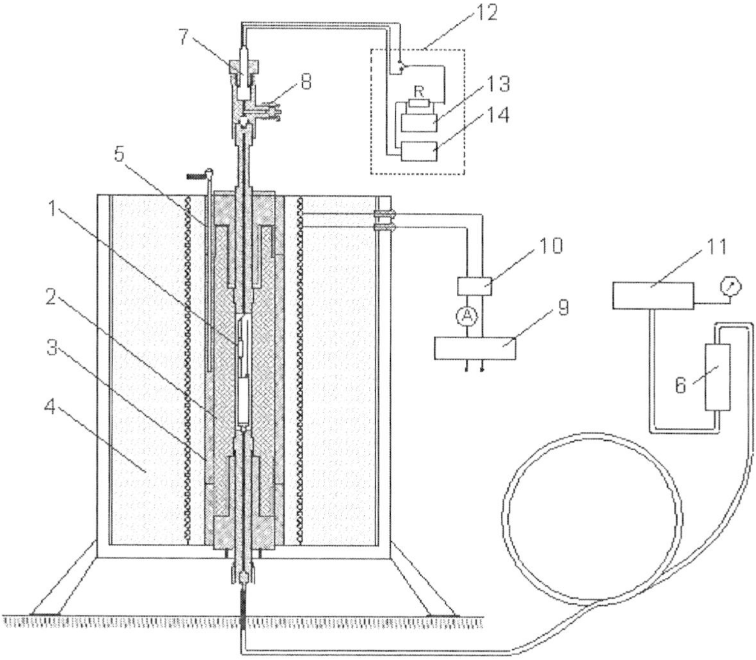

Figure 2a. High temperature and high pressure capillary viscometric system (Semenyuk et al. [49]). Titanium viscometer-1; high-pressure autoclave-2; copper block-3; cylindrical electro-heater-4; PRT-5; separating vessel-6; high-pressure electro-input unit-7; removal (withdrawal)-8; electronic voltage stabilizer (П71М)-9; auto-transformer-10; dead-weight pressure gauge (MP-2500)-11; indicated circuit-12; electronic oscillograph-13; audio-frequency oscillator-14.

In Eq. (3) the viscosity obtained is proportional to the fourth power of the capillary radius. Hence a very accurate value of the radius is required. The inner surface of the capillary walls is perfectly polished with powders of successively smaller grain size (1 to 40 nm). The average capillary radius is measured using a weighing technique or/and by calibration (relative method, see Abdulagatov and Azizov [46]) from the viscosity of a standard fluid (pure water) with well-known viscosity values (IAPWS formulation, Kestin et al. [52]). The correction for capillary radius, due to the meniscus curvature (surface forces in a capillary), is made. Typical values (obtained by Abdulagatov and Azizov [46]) of the capillary radius determined with both weighing and by the calibration techniques are 0.15091 mm and 0.15048 mm, respectively, resulting in a value for R^4 of $(51.27 \pm 0.59)\mathrm{x}10^{-5}$ mm^4. The length, L, of the capillary is measured with a microscope with an uncertainty of ±0.005 mm. Typically the final value of the capillary length (obtained by Abdulagatov and Azizov [46]), is 540.324 ± 0.005 mm. Flow time measurements are usually made electronically and the uncertainty in this value is never more than 0.1s (0.5 %).

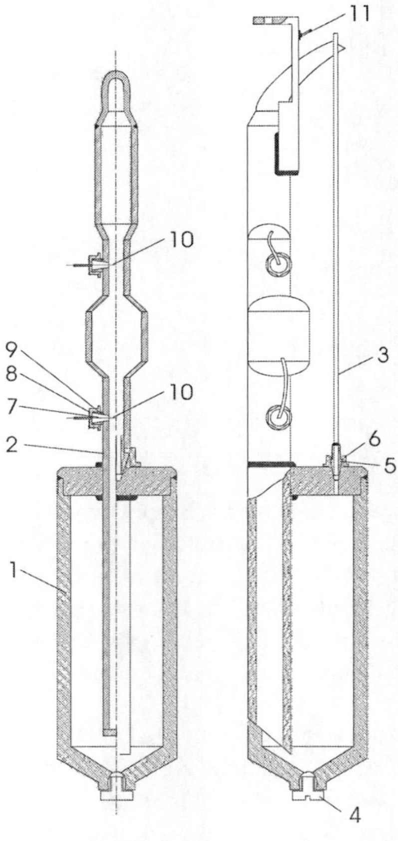

Figure 2b. Construction of the titanium viscometer (Semenyuk et al., [49]). Cylindrical bulb-1; shaped (figured) tube-2; titanium capillary-3; thread lock (stopper)-4; conical packing-5; adapter nuts-6 and 8; insulators-7; conical pockets-9; platinum contacts-10; common contact-11.

To assess the uncertainty of the technique, the uncertainty of each quantity that enters into Eq. (3) must also be taken into consideration. After a very careful analysis of the uncertainty of these quantities (Abdulagatov and Azizov [46]), it is estimated that the combined relative uncertainty in measuring the viscosity (up to the highest temperature of 575 K) is 1.5 %. The uncertainty in the pressure measurements is 0.05 %.

A very similar viscometric apparatus was developed by other authors in the works [47-51] to measure the viscosity of aqueous systems. They used nickel, glass, and platinum viscometers. The uncertainty of the measured values of viscosity quoted is 1.5 % for the glass viscometer and 1.0 to 1.5 % for the nickel viscometer. The internal radius and length of the metallic capillary was 0.15 mm and 13.5 cm, respectively. The viscometric apparatus used by Semenyuk et al. [49] schematically shown in Figure 2a,b. Figure 2b shows the details of the construction of titanium viscometer.

Recently we developed new apparatus (glass capillary viscometer) to measure the viscosity of liquid substances. This apparatus was use to measure the viscosity of fruit juices (Magerramov et al. [14,15]). The apparatus used to the viscosity measurements is schematically shown in Figure 3a,b. Details of the apparatus and procedure were described elsewhere [14,15,33-37]. The apparatus for the viscosity measurement consisted of (see Figure 3a) a high pressure vessel -1, hydraulic press -2, pinching vessel-3, and glass viscometer -4. Main part of the apparatus is glass viscometer-4 which was located in the high temperature and high pressure autoclave with conical packing-5. The high pressure vessel-1 was made from stainless steel (1X18H9T). In order to equalization and maintain homogeneous temperature during the measurements the steel vessel was covered on the cooper block- 8. The high pressure viscometric vessel was supplied with two semi-axis -7 to maintain on the frame and for the rotation around the horizontal axis. The grooves were milled and drilled the well in the copper block for the resistance thermometer and controlling differential chromel–copel thermocouples. One of the semi-axis of the high pressure vessel was supplied with stopper mechanism to fix of the vessel strict vertical position by reading control pointer and scale, maintained on the end of second semi-axis -7. The high pressure autoclave was immersed in a thermostat-6 which was made from the two semispherical iron sheets. The gap between the sheets was filled with asbestos isolator. Two electrical heaters were wound around the surface of the copper block. To create and measure of the pressure, the autoclave was connected with a dead-weight pressure gauge (MP-600) by means of separating vessel. The temperature of the juices was measured with a PRT-10 (platinum résistance thermometer, 10 Ω) using potentiometer circuit. The high-pressure electro-output units were used to connect the viscometer with time of fluid flowing measuring circuit. The high-pressure electro-output unit consists of pinching cone with lock nut, pinching holder -9 and connecting tube. High pressure vessel was immersed in thermostat -6 and maintained on the supports of frame and can rotate around horizontal axis-7. Viscometer was made from refractory glass "supremacs". A construction of the capillary viscometer is shown in Figure 3b. The viscometer consists of a lower bulb. The connecting tube with oval shape is located inside the lower bulb. The lower end of the connecting tube is 2-3 mm above the bottom of the bulb. The upper end of the connecting tube goes to measuring and preliminary bulb. The capillary was welded to the side of the preliminary bulb parallel to the vertical axis of the viscometer. For the centering of the viscometer in the high pressure vessel the lower bulb supplied with shoulders and with funnel for the mercury filling. In order to contact with mercury the platinum wires were seal to 3 contacts– two on the inlet of the measuring bulb and other one on the lower bulb. The platinum contacts of the viscometer were welded to insulated nichrome wires which were connected to the outs of the pressure vessel by an electrically insulated feed-through. The viscometer was suspended to packing cone in the high-pressure vessel by using tag. The tag was supplied with four branches to supporting electro-output units of the viscometer.

Operating Procedure

Initially at vertical position mercury is in the lower bulb. When the high-pressure viscometric vessel is turned by an angle of 90° the viscometer is in a horizontal position and the mercury spills over the whole viscometer volume. When the viscometer returns to its initial vertical position, the level of mercury will be higher than the upper contact. Due to the difference of the mercury levels in the viscometer, flow of the fluid through the capillary takes place. While lowering, the mercury successively disconnects at the inlet and the outlet of the measuring bulb and the flow time is fixed.

The time of fluid flowing through the capillary τ was measured automatically by a frequency-meter with an uncertainty of 0.01 s. The upper contact of the viscometer was connected to brand "start" and the middle to brand "stop". The measurement of the flow time was repeated 5-6 times for each temperature and pressure in order to confirm the reproducibility of the results. The flow time for the investigated juices was about 45 s at high temperatures (120-130 °C) and 1000 s at low temperatures (at room temperature).

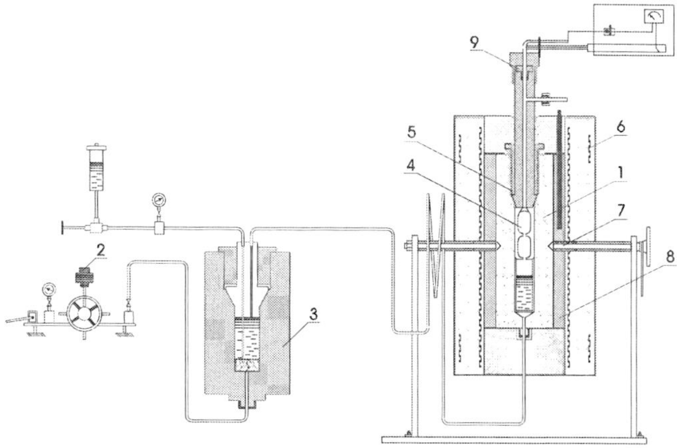

Figure 3a. Schematic diagram of the experimental apparatus for viscosity measurements at high temperatures and high pressures (Magerramov et al. [14,15]). 1 – high-pressure vessel; 2- hydraulic press; 3- pinching vessel; 4-glass viscometer; 5-conical packing; 6- thermostat; semi-axis-7; 8-copper block; 9-packing cartridge.

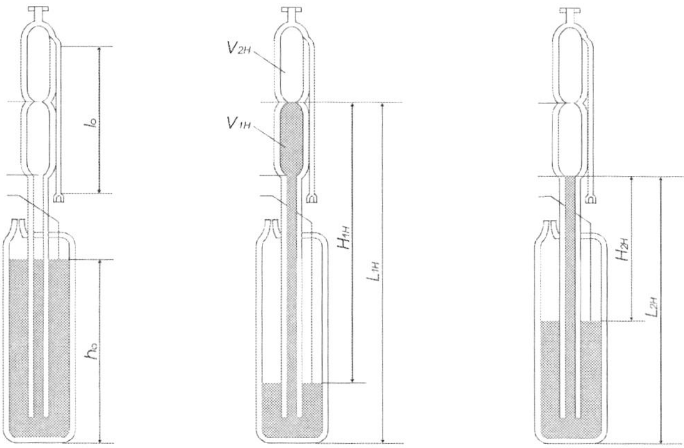

Figure 3b. Construction and geometric characteristics of the high-pressure capillary viscometer (Magerramov et al. [14,15]). V_{1H} and V_{2H} are volumes of the measuring and preliminary bulbs, respectively; H_{1H} and H_{2H} are the mercury level drop at the beginning and the end of the flow; L_{1H} and L_{2H} are the height of the mercury column at the beginning and at the end of the flow, respectively; h_0 is the height of the mercury in the lower bulb; l_0 is the capillary length.

The viscosity was obtained from the measured quantities (r_0^4, H_0, h_0, L_0, l_0, V_{10}, τ, ρ_{Hg}, ρ_{Hg0}, T, P, ρ, and m): geometric sizes of the viscometer; radius and the length of the capillary; the volumes of the measuring and preliminary vessels, the difference of the mercury levels, and total length of the mercury column in the beginning and the end of the flow. The geometrical size of the viscometer was determined using the method (mercury –weighing method) described in (Abdulagatov and Rasulov [37]). The viscometer constants were determined using microscope (MIR) and cathetometer. The radius was determined by weighing and relative methods. The dimensions and geometric characteristics of the viscometer are: V_{1H}=1.675×10^{-6} m^3; V_{2H} =1.430×10^{-6} m^3; r_0=12.58×10^{-5} m; l_0=5.942×10^{-2} m; L_0=9.572×10^{-2} m; H_0 =5.154×10^{-2} m; h_0=5.052×10^{-2} m.

Working Equation

The measurements of the viscosity is based on Poiseuille's law which relates viscosity η to the rate $v = V/\tau$ of fluid flow through a capillary tube

$$\eta = \frac{\pi r^4 \Delta P \tau}{8Vl}, \tag{6}$$

where ΔP is the pressure drop ($\Delta P = P_{in} - P_{out}$, where P_{in} is the inlet pressure, P_{out} is the outlet pressure), r is the inner radius of the capillary, V is the volume of the fluid flowing through the capillary for the time τ, l is the capillary tube length, τ is the time of flow. After corrections (which take into account the acceleration of a fluid at the inlet and outlet; corrections for the effect of thermal expansion of the mercury and glass; corrections for the changing of the mercury level in the viscometer with temperature and pressure) the variation of the geometrical sizes of the capillary, mercury and sample densities at the experimental conditions were varied with temperature and pressure; the final working equation for the viscosity can be written as

$$\eta = A\left(C_t - a\frac{\rho_{Hg0}}{\rho_{Hg}}\right)(\rho_{Hg} - \rho)\tau - B_t\frac{\rho}{\tau}, \tag{7}$$

where the viscometer constants are

$$A=\frac{\pi g r_0^4 H_0}{8V_{10}l_0},\quad B_t=\frac{mV_{10}}{8\pi l_0}\left(1+2\alpha\Delta t\right),\quad C_t=\left(1+\frac{h_0+3\alpha\Delta t L_0}{H_0}\right),$$

and

$$a=\frac{h_0}{H_0},\tag{8}$$

where V_{10} is the measuring volume, ρ_{Hg} is the density of mercury at the experimental conditions (at experimental T and P); ρ is the density of the liquid under study at the experimental conditions, ρ_{Hg0} is the density of mercury at room temperature; H_0 is the average mercury level drop; L_0 is the average height of the column at the flowing process; h_0 is the height of the column in the lower vessel at the initial position; $\alpha = 4.31\times10^{-6}$ K^{-1} is the linear expansion coefficient of the capillary material; r_0 is the capillary radius; l_0 is the length of capillary; Δt is the temperature difference between experimental temperature and room temperature; m=1.12 is a constant introduced to take account of the shape of the capillary ends (correction factor).

The values of the parameters (viscometer constants) A, B_t, and C_t in Eq. (8) can be also determined by means of a calibration procedure from the viscosity of a reference fluid (for example, pure water IAPWS standard data, Kestin et al. [52]) with well-known viscosity values. In the present study the values of parameters A and B_t were determined by calibration (5.895×10^{-10} m^2 s^{-2} and 1.2498×10^{-6} m^2, respectively) on pure water and by the geometric characteristics of the apparatus (5.849×10^{-10} m^2 and 1.2571×10^{-6} m^2, respectively). The values of the parameters a and C_t calculated with geometric characteristics are 0.97497 and 1.9808, respectively. The change in the capillary radius (r_0), length of capillary (l_0), and in the measuring volume (V_{10}), therefore, and in the values of viscometer constants A due to pressure, was considered negligible due to the low volume compressibility of the capillary material (stainless steel 1X18H9T). The effect of pressure on geometrical characteristics of the cell is also negligibly small in the pressure range of the experiments since the entire viscometric capillary was under pressure.

Uncertainty of the Measurements

The uncertainty analysis was based upon the ISO Uncertainty Guide (IOS 4787 [53]) and Coleman and Steele [54]). The uncertainties reported in this paper are expanded uncertainties at 95 % confidence level with a coverage factor of k=2. The accuracy of the viscosity measurements strongly depends on the uncertainty of each individual measurement involved in the overall determination. In this method the measurement of the following basic quantities are needed: r_0^4, H_0, h_0, L_0, l_0, V_{10}, τ, ρ_{Hg}, ρ_{Hg0}, T, P, ρ, and m. The accuracy of the viscosity measurements was assessed by analyzing the sensitivity of Eq. (7) to the experimental uncertainties of the measured quantities from which the viscosity is determined. The uncertainties all of the measured quantities are given in Table 3. The expanded total uncertainty ($\delta\eta/\eta$) with a coverage factor of k =2 at 95 % confidence for the viscosity measurement in this method is

$$\frac{\delta\eta}{\eta} = \frac{1}{\eta}\sqrt{\left(\frac{\partial\eta}{\partial R^4}\right)^2 S_{R^4}^2 + \left(\frac{\partial\eta}{\partial\Delta H_0}\right)^2 S_{\Delta H_0}^2 + \cdots + \left(\frac{\partial\eta}{\partial x}\right)^2 S_X^2}\,, \qquad (9)$$

where $S_{R^4}^2$, $S_{\Delta H_0}^2$, ..., and S_X^2 are the root-mean-square deviations of R^4, ΔH_0, ..., and m, respectively. The systematic uncertainties of the viscosity measurements can be estimated from the equation

$$Q_\eta = \frac{1}{2}\left[\left(\frac{\partial^2\eta}{(\partial R^4)^2}\right)S_{R^4}^2 + \left(\frac{\partial^2\eta}{\partial(\Delta H_0)^2}\right)S_{\Delta H_0}^2 + \cdots + \left(\frac{\partial^2\eta}{\partial x^2}\right)S_X^2\right]. \qquad (10)$$

The uncertainty all of the measured quantities was determined as $S = \sqrt{\sum_{i=2}^{n}\left(x_i - \bar{x}\right)^2 / n(n-1)}$, where x_i is the measuring quantity; $\bar{x}$ is the mean value; n is the number of measurements. Based on the detailed analysis of all sources of uncertainties likely to affect the determination of viscosity with the present apparatus, the combined maximum relative uncertainty $\delta\eta/\eta$ in measuring the viscosity was 1.5 %. The relative systematic uncertainties Q_η/η was 0.001 %.

Table 3. The uncertainty of the measured quantities.

No.	Measured quantities	Uncertainty
1	Height of the mercury column in the viscometer, L_0, m	9.572×10-2
2	Level of the mercury in the lower vessel, h_0, m	5.248×10-2
3	Mercury level drop at the beginning of flow, H_0, m	5.154×10-2
4	Measuring volume, V_{10}, m3	1.675×10-6
5	Length of the capillary, l_0, m	5.942×10-2
6	Thermal expansion coefficient of the viscometer, α, K-1	4.31×10-7
7	Uncertainty in capillary radius determination, m	2.16×10-7
8	Uncertainty in capillary length determination, m	5.5×10-8
9	Uncertainty in measuring volume of viscometer, m3	1.65×10-10
10	Uncertainty in height of the mercury column, m	5.0×10-2
11	Uncertainty in temperature measurements, K	2.5×10-2
12	Uncertainty in fluid flowing time measurement, s	1.0×10-2
13	Relative uncertainty in liquid density measurements,	2.0×10-4
14	Relative uncertainty in pressure measurements	6.0×10-5
15	Relative systematic root-mean-square uncertainty in viscosity measurement	1.6 %
16	Random root-mean-square uncertainty in viscosity measurement	0.1 %
17	Total root-mean-square uncertainty in viscosity measurement	1.7 %

2.1.2. Oscillating-Disk Technique

Oscillating-disk viscometers consist of an axially symmetric disk placed between two fixed parallel plates and suspended from a torsion wire, so that it performs oscillations in the fluid about its axis of symmetry. Such a viscometer has been used by several researchers as an instrument for viscosity measurements (see, for example [55-62]). The theoretical and experimental problems of this technique have been discussed in detail elsewhere [61,63-67].

2.1.2.1. Theoretical

The theoretical aspects of the technique are described in the works [65-67]. The characteristic equation which relates the motion of an oscillating body to that of the fluid is [67]

$$(s+\Delta_0)^2+1+D(s)=0\,, \tag{11}$$

where Δ_0 is the logarithmic decrement in a vacuum. Here, $D(s)$ is the torque on the body, calculated from solutions for the fluid flow close to the oscillating body. It is therefore the torque $D(s)$ that is characteristic of the particular type of oscillating viscometer employed.

Apart from roots that are associated with the initial oscillatory behavior of the body, the roots associated with the long-time behavior can be written in the form

$$s=\pm(i-\Delta)/\theta \tag{12}$$

where i is the imaginary unit and $\theta=T/T_0$. Here, T is the period of oscillation in the fluid and the subscript zero denotes the same quantity in vacuo. The function $D(s)$ can be expressed as

$$D(s)=\frac{s\,M(s)}{I\omega_0^2\,[s\alpha(s)-\alpha_0]}, \tag{13}$$

where $M(s)$ is the Laplace transform of the torque function $M(\tau)$ (torque exerted by the fluid on the body), $\tau=\omega_0 t$ is a dimensionless time, $\omega_0=2\pi T_0$ is the angular frequency of oscillation in a vacuum, I is the moment of inertia of the oscillating system. The symbol, α_0, denotes the initial angular displacement, while $\alpha(s)$ is the Laplace transform of the angular displacement $\alpha(\tau)$ of an axially symmetric body, defined as

$$\alpha(\tau) = \frac{\alpha_0}{2\pi i} \int_C \left\{ \frac{1}{s} - \frac{1+\Delta_0^2}{s\left[(s+\Delta_0)^2 + 1 + D(s)\right]} \right\} e^{s\tau} ds \tag{14}$$

or

$$\alpha(\tau) = \alpha_0 e^{-\Delta\theta\tau} \cos(\theta\tau) . \tag{15}$$

During an experiment, the logarithmic decrement Δ and the period T, are determined from the motion of the body. The form of the torque functions D(s) for various oscillating-body viscometers are discussed in the literature [66, 68-71].

Before proceeding to examine different forms of the function D(s), it is important to introduce a natural length scale that appears in oscillatory systems in general, namely the boundary-layer thickness, δ, defined as

$$\delta = \left(\frac{\eta}{\rho\,\omega_0} \right)^{1/2} . \tag{16}$$

As it will be shown in the following paragraphs, this natural length scale is important in the selection of the dimensions of oscillating-disk viscometers.

In the particular case of viscometers consisting of a disk of thickness d and radius R, oscillating between two fixed horizontal plates, and provided $\delta >> 2b+d$ (where b is the separation between the disk and one of the fixed horizontal plates), the solution for D(s), is [61]

$$D(s) = \frac{\pi\rho R^4 \delta}{I} s^{3/2} \left[\coth\left(\frac{b}{\delta} s^{1/2} \right) + \frac{(C_N - 1)\delta}{b} s^{1/2} \right], \tag{17}$$

where C_N is calculated from the viscometer geometry [72].

For the oscillating-disk viscometers consisting of a thick disk oscillating in an infinite fluid (i.e. a free thick disk), and provided that $\delta << d/2$ and $R >> \delta$, the analytical solution for $D(s)$ is [61]

$$D(s) = \frac{\pi \rho R^4 \delta}{I} s^{3/2} \left[1 + \frac{2d}{R} \frac{K_2(\xi_0 s^{1/2})}{K_1(\xi_0 s^{1/2})} + \varepsilon(s) \right], \qquad (18)$$

where K_1 and K_2 are the modified Bessel functions of order 1 and 2, and $\xi_0 = R/\delta$. The edge-effects $\varepsilon(s)$ for this type of geometry are given by

$$\varepsilon(s) = \frac{16}{3\pi} \left[\frac{4\pi}{3\sqrt{3}} - 1 \right] \frac{1}{\xi_0 s^{1/2}} + \frac{17}{9} \frac{1}{\xi_0^2 s} + \ldots \qquad (19)$$

The solutions represented by Eqs. (16) and (19) are valid for oscillations of small amplitude.

For all disks the general torque function can be expressed (Nieuwoudt et al. [69]) in the form of

$$D_k(s) = \frac{\rho}{\bar{\rho}} s^{3/2} \left[A_k \xi_0^{-1} + B_k s^{-1/2} \xi_0^{-2} + C_k s^{-1} \xi_0^{-3} + D_k s^{-3/2} \xi_0^{-4} + Q_k(s) \right], \qquad (20)$$

where k denotes the type of configuration and $\bar{\rho}$ is the effective density of the oscillating body. The term A_k arises from the torque on the flat and cylindrical sides of the disk, while the high-order terms contain the effects of the sharp edges at the rims of the disk. The values of the coefficients A_k, B_k, C_k and D_k appearing in Eq. (20) for various oscillating-body viscometers, can be found in literature [41].

2.1.2.2. Experimental

During the last fifty years, many designs of oscillating-disk viscometers, operating in a relative or in an absolute way, have been reported and their theory has been developed in the works [56,60,61,63-67,73-75]. However, due to design limitations, absolute measurements of the viscosity of aqueous solutions with an oscillating-disk viscometer are almost impossible. Kestin and coworkers [56,58,61] employed relative methods, based upon the development of previous formulations [64-67] in order to measure the viscosity of aqueous solutions at temperatures up to 200 °C and pressures up to 30 MPa. Calibration of the viscometer involves measurement of the viscosity of distilled water, whose viscosity is well known, in order to determine the dependence of the edge-correction factor C upon the boundary-layer thickness δ. Calibration was performed over the temperature range from 25 to 150 °C, which corresponds to a boundary-layer range from 0.73 to 1.53 mm.

The oscillating-disc viscometer used by Kestin et al. [56,57,60], Grimes et al. [74], and Kestin and Shankland [75] to measure the viscosity for aqueous solutions at high temperatures and high pressures, is schematically shown in Figure 4.

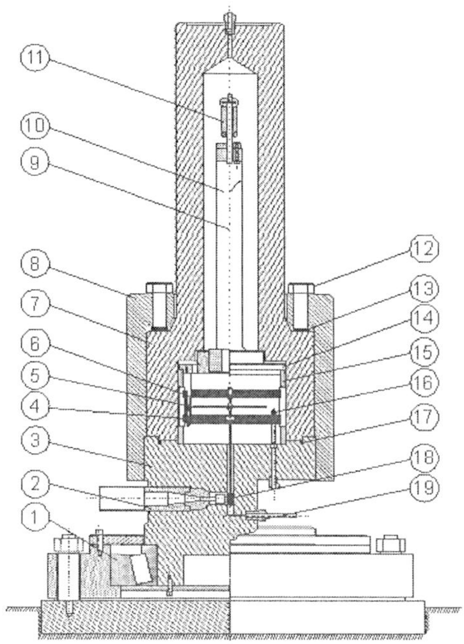

Figure 4. Cross-section view of the viscometer by Kestin *et al.* [56]. Bearing for rotating viscometer-1; window-2; solid lower part of the viscometer-3; fixed plates-4,6; disk-5; hollow upper part of the viscometer-7; buttress-threaded stainless steel cap-8; Pt/W thin strand-9; alumina column-10; suspension assembly-11; ten bolts-12; pressure ring-13; suspension mounting plate-14; thick-walled cylinder-15; thermocouple-16; O-ring seal-17; mirror-18; filling connection-19.

The viscosity body is divided into a solid lower part -3 and hollow upper part- 7 forming the viscometer cavity, effectively sealed against an internal pressure in excess of 30 MPa. The oscillating-disk -5 is suspended between two Hastelloy fixed plates -4 and -6 by means of a stress relieved 92 Pt thin strand -9. The upper end of the wire is gripped at the top of the suspension cartridge. The top suspension assembly -11 is designed to permit controlled adjustments in the elevation of the disk as well as in the angular orientation of the oscillating system. Two side windows at the lower part of the instrument provide access to the disk and the fixed plates.

2.1.2.3. Working Equation and Uncertainty

As already mentioned, the value of the viscosity (obtained in a relative way), is derived from measurements of the logarithmic decrement Δ and the period of oscillation T. The working equation for the viscometer reduces to [56,58]

$$\frac{2}{\theta}\left(\frac{\Delta}{\theta}-\Delta_0\right)-C\frac{\pi\rho R^4\delta}{I}\left[H_1K_2+H_2K_1+\frac{2d}{R}\left(H_1+\frac{3\delta}{2R\theta}\right)\right]=0, \tag{21}$$

where, θ is the ratio of the period of oscillation in the liquid to that in vacuum, R is the radius of the disk, I its moment of inertia, and δ is a measure of the boundary layer thickness (Eq. (16), $\delta=\sqrt{\eta/(\rho\,\omega_0)}$). Parameter C is an edge-correction factor depended on the dimensionless boundary-layer thickness (δ/b), determined by calibration. Finally, the functions $H_{1,2}$ and $K_{1,2}$ are defined by

$$H_{1,2}=\left(1\pm1.5\Delta-0.375\Delta^2\right)/\sqrt{2}\theta^{3/2}, \tag{22}$$

and $$K_1=\frac{\sin Y}{\cosh X-\cos Y},\quad K_2=\frac{\sinh X}{\cosh X-\cos Y}, \tag{23}$$

with $$X,Y=\frac{2}{\sqrt{2\theta}}\left[1\pm0.5\Delta+0.125\Delta^2\right]\frac{b}{\delta}. \tag{24}$$

Eq. (21) is first employed in combination with viscosity measurements of a liquid of "known" viscosity, in order to obtain the edge-correction factor $C($

δ). Consequently having obtained this factor, it is applied to measure the viscosity of other liquids from measured values of Δ and θ .

The precision of the viscometer depends both upon the accuracy of the mathematical model describing its operation and on the precision of each individual measurement involved in the overall determination. In this method the measurement of the following basic quantities are needed: temperature *T*, pressure *P*, a series of the times, and the period of oscillation.

The temperature inside the viscometer is usually measured by means of a Pt/Pt-Rh thermocouple with an uncertainty of 0.05 K. The uncertainty in pressure measurements is 50 kPa. The time intervals were measured with the aid of digital timers and a photo-electric circuit. The accuracy of the time interval measurements is 1 ms; this translates into a logarithmic decrement, known to within 0.1 % and a period that is known to within 0.001 %. On the basis of these values the estimated uncertainty of the viscosity measurements does not exceed 0.4 %. Kestin et al. [60] numerically evaluated the sensitivity of the viscometer working equation to uncertainties in density. They found that at room temperatures, the viscosity is insensitive to any uncertainty in density, while at 200 °C, the relative uncertainty in density is amplified by a factor of approximately 2 in viscosity.

2.1.3. Falling-body Technique

This method is usually employed to measure the viscosity of liquids and high-dense gases. This method is characterized by high uncertainty (3-4 %), but has some advantages for measurements at high pressures. The theoretical and experimental problems of this technique have been discussed by Kawata et al. [76].

2.1.3.1. Theoretical

The method is based on Stoke's law, according to which, the force, *W*, exerted by the liquid on a sphere of radius, *r*, falling in an infinite homogeneous liquid at a constant velocity , is related to viscosity η of the liquid via the equation W = 6¾Ár v . This relation is valid for very small Reynolds numbers, *Re*. At greater Reynolds numbers, this relation should be corrected for the effect of walls. Stoke's law assumes that the following conditions are satisfied, during the free-fall under gravity of a sphere through the liquid of interest (Wakeham et al. eds. [41]): (a) the motion of the sphere is sufficiently slow; (b) the liquid is of infinite extent; (c) there is no slip between the liquid and the surface of the sphere; (d) the sphere is rigid; (e) the liquid is incompressible; and (f) the liquid is Newtonian.

For falling bodies of various shapes (a cylinder, a cylinder with semispherical ends, or an arrow-like body), the viscosity is determined by measuring the time t of fall through a fixed distance at a constant velocity as

$$\eta = C\left(\rho_{\mathrm{b}} - \rho_{\mathrm{l}}\right)t\,, \tag{25}$$

where C is the instrument's constant (function of the dimensions and the design of the apparatus), ρ_b is the density of the falling body, and ρ_l is the density of the liquid studied. The value of the instrument's constant, C, is usually determined by a calibrating procedure, from the viscosity of a standard fluid (usually pure water) with well-known viscosity values (IAPWS formulation, Kestin et al., [52]).

For a free-fall of a sphere of radius r through a distance h, in a liquid enclosed in a tube of radius R, the viscosity equation is derived as

$$\eta = \frac{2r^2 g}{9h}(\rho_{\mathrm{b}} - \rho_{\mathrm{l}})t\, f_w\,. \tag{26}$$

In the above equation f_W is the correction factor for the wall effect. The correction factor f_W has been studied theoretically and experimentally [76-79]. Faxen [77] expressed the factor f_W as a function of the Reynolds number, Re, and the ratio (r/R)

$$f_w = 1 - 2.104(r/R) + 2.09(r/R)^3 - 0.95(r/R)^5\,. \tag{27}$$

Stoke's law is applicable provided the Reynolds number, Re, is much less than unity. Oseen [80] derived the following working equation for a sphere moving with a velocity v, which is applicable for a wide range of Re number

$$\eta = \frac{2r^2 g}{9v(1 + 3Re/16)}(\rho_{\mathrm{b}} - \rho_{\mathrm{l}}) \text{ for } Re = \frac{2Rv\rho_{\mathrm{l}}}{\eta} < 1 \tag{28}$$

There are several modifications of the Stoke's law for a wide range of *Re* numbers (see, for example, [79-83]). Other corrections to the working equation of this technique, such as the effect of the fall-tube ends, the fall-tube dimensions, and the selection of spherical material, are well documented in literature [84-86].

For a cylinder of mass *m*, radius r_1 and length L_b, falling under the influence of gravity a distance *h* along the axis of a tube of radius r_2, the viscosity of the liquid can be expressed as

$$\eta = \frac{(\rho_b - \rho_1)}{A\rho_b} t \tag{29}$$

where $$A = \frac{2\pi L_b h}{mg\left[\ln(r_2 / r_1) - [(r_2^2 - r_1^2)/(r_2^2 + r_1^2)]\right]}. \tag{30}$$

In practice the value of the instrument's constant *A* is determined by a calibration procedure employing fluids of well-known viscosity (pure water, for example). This method was used to measure the viscosity of water, alcohols, and their mixtures [87,88].

2.1.3.2. Experimental

Melikov [89-91] employed the falling-body technique to measure the viscosity of aqueous solutions at temperatures up to 300 °C and pressures up to 30 MPa. The main part of the apparatus is a viscometric tube - the fall-tube - (see Figure 5a-c), with 40 mm OD, 10.33 mm ID, and a length of 500 mm, made out of stainless steel. The working section of the viscometric tube is 400 mm (see Figure 5b). The inner surface of the tube walls is perfectly polished with powders of successively smaller grain size (1 to 40 nm). On the outside surface of the fall-tube (50 mm from the ends), the two inductance coils (electromagnetic transducers) are mounted, connected to potentiometers. When the falling-body passes the inlet and outlet of the fall-tube, an induced signal triggers the appropriate timing circuitry. The cylindrical falling body with a diameter of 10.214 mm and a density ρ_b = 6.290 g·cm^{-3}, is made out of magnetic steel. The viscometer is located in a liquid thermostat with 7 heaters and can be rotated upside down in order to let the ball return to its starting position.

To perform accurate viscosity measurements with the falling-body technique, various corrections (fall-tube dimensions, effect of fall-tube ends, terminal velocity, falling-body shape, position of the fall-tube) ought still to be considered (Wakeham et al. eds. [41]).

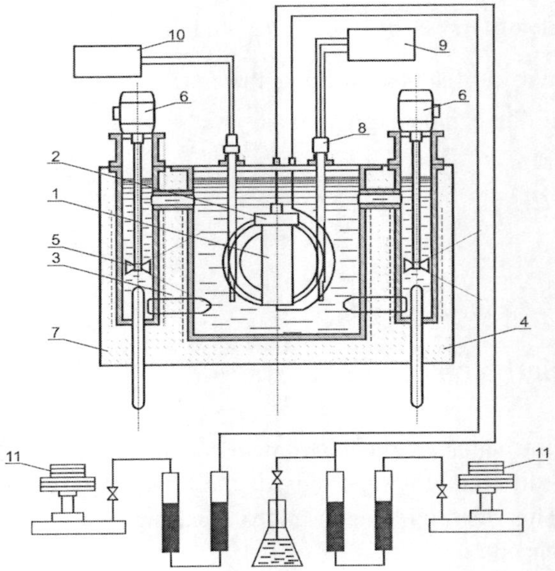

(a)

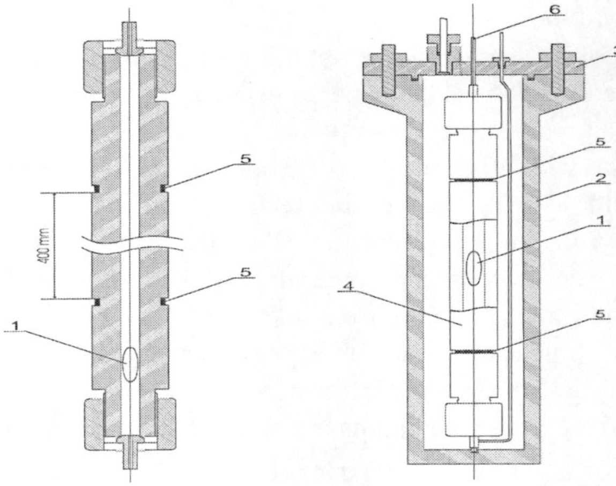

(b)

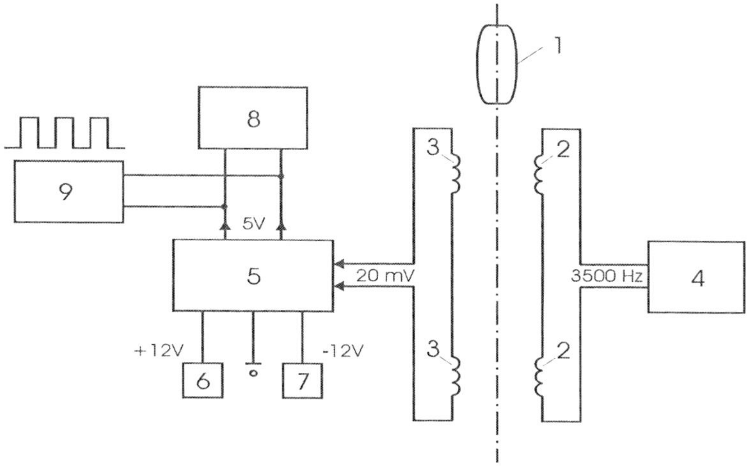

(c)

Figure 5a-c. Schematic diagram of a falling-body viscometer (Melikov [89]). Experimental apparatus (5a), fall-tube construction (5b), and fall-time measure unit (5c). (a): 1-high-pressure autoclave; 2-cover of the high-pressure vessel; 3-rotary encoder; 4-liquid thermostat; 5-mixer; 6-circulating pump; 7-thermostating liquid; 8-PRT-10; 9-patentiometer (P363/1); 10-high precise temperature regulator (HPTR-3); 11-manometer, (b): 1- falling-body; 2-high pressure autoclave; 3- cover of the high-pressure vessel; 4- fall-tube; 5- inductance coils (electromagnetic transducers); 6-filling tube, (c): 1- falling-body; 2-primary winding of the transducer; 3-secondary winding of the transducer; 4- generator of the signals; 5-amplifier; 6,7-direct current power supply unit; 8- frequency meter; 9-oscillograph.

2.1.3.3. Uncertainty

The uncertainty of the viscosity values is directly connected to the uncertainty associated with several other factors, including the measurement of the dimensions, mass and density of the falling body, the radius of the fall tube, the distance traveled and time taken, the density of the falling body and that of the liquid. These uncertainties are connected through the corresponding working equation. In particular the radius of the falling body and that of the tube are of primary importance as they strongly affect the viscosity. For the instrument described here (Melikov [89-91]), the uncertainty in the radius of the falling cylinder and the tube is 0.001% in both cases. Density of falling body and liquid were measured with an uncertainty of 0.002% and 0.0052% respectively. The temperature and the pressure were measured with an uncertainty of 0.1 K and 0.005 MPa. Hence, the resulting total uncertainty in the viscosity measurements was about 3.3 %. It should also be pointed out that as a test of the operation of his instrument, Melikov [89-91] measured the viscosity of pure water at pressures between 0.1 and 40 MPa and over the temperature range from 298 to 573 K. His values agreed with the IAPWS formulation values (Kestin et al. [52]) within 1.5 %.

Chapter 3

Discussion of Available Experimental Viscosity Data for Fruit Juices

In Table 2, a summary of all viscosity measurements, to our knowledge, is presented for liquid foods. In the same table, for every fruit juice, the authors, the concentration and the temperature ranges, and the experimental method employed are shown. The tables provide (to our knowledge) the largest and most useful collection of viscosity data at the present time for the liquid foods. As one can see from this Table 2, most available experimental viscosity data sets for liquid foods basically cover a low temperature range (below 60 °C) and limited concentration range. All previous viscosity measurements for liquid foods were performed at atmospheric pressure. Very restricted (about 25 datasets) measurements are available in the literatures at high temperatures (see Table 2) and no data reported at high pressures, except the data reported by the present authors (Magerramov et al. [14,15]) for four fruit (pomegranate, pear, lemon, and tangerine) juices. Majority of viscosity measurements of fruit juices have been made using the concentric cylinder (Haake Rotovisco RV 12), capillary tube, falling sphere, rotational viscometers, and controlled stress rheometer (RheStress RS1) techniques. Most authors used the measured viscosity data to calculate the values of parameters (flow activation energy E_a and η_0) in the Arrhenius viscosity equation. The effect of temperature and concentration on viscosity of fruit juices was study using the various type models (Arrhenius-type equation, polynomial, exponential type, power law and their various combinations).

Figures 6 to 10 shows the typical dependence of the viscosity of fruit juices on temperature, pressure, and concentration. The following four sections 4.1 to 4.4 will provide a brief outline of the available schemes describing the temperature, the pressure and the concentration dependence of the viscosity of liquids and fruit juices. These schemes will be employed to the viscosity of fruit juices in order to perform examine the predictive capability and accuracy of the models and their applicability to the viscosity of fruit juices.

Below we briefly analyzed the reported viscosity datasets for fruit juices. The viscosity measurements for four fruit (pomegranate, pear, tangerine, and lemon) juices have been performed in the temperature range from 292 to 403 K at pressures up to 10 MPa by using the method described above (section 2.1.1.2). The concentration range was between 11 and 45 °Brix. All experimental viscosity data were obtained as a function of temperature for the several fixed concentrations at three isobars (0.1, 5, and 10 MPa). The experimental temperature, viscosity, pressure, and concentration values for the pomegranate, pear, tangerine, and lemon juices are presented in Appendix. The comparison of the data with the values reported by other authors is shown in Figures 11 to 14.

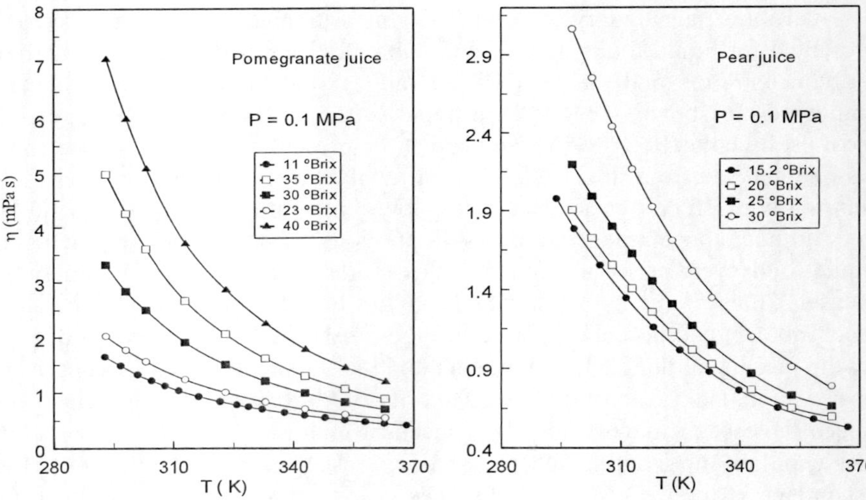

Figure 6. Measured values of viscosity η of pomegranate and pear juices as a function of temperature T along the various constant concentrations. Symbols – experimental values (Magerramov et al. [14]), solid lines calculated with correlation model (Eq.32).

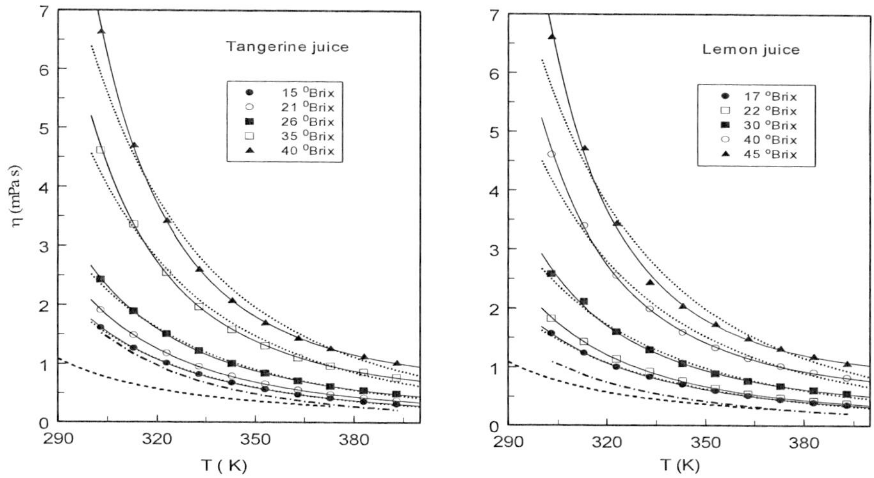

Figure 7. Measured values of viscosity η of tangerine and lemon juices as a function of temperature T along the various constant concentrations. (------), pure water calculated with IAPWS formulation (Kestin et al. [52]); (———), calculated with Eq. (32); (-·· - · · · -), Alvarado and Romero [92] for x=9.4 °Brix (tangerine juice) and x=8.4 °Brix (for lemon juice).

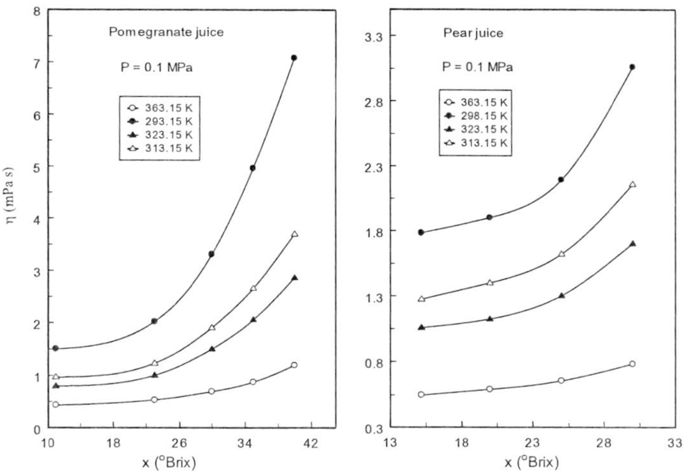

Figure 8. Measured values of viscosity η of pomegranate and pear juices as a function of concentration along the various selected isotherms (Magerramov et al. [14]).

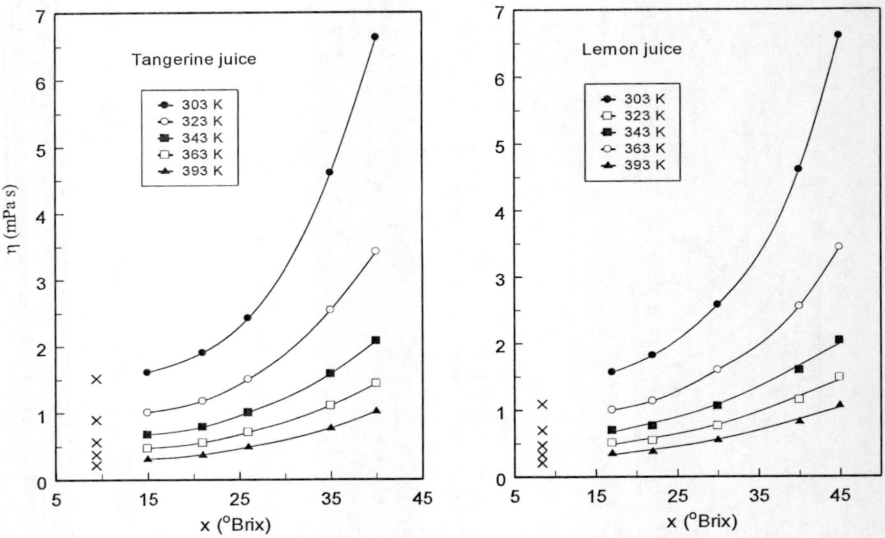

Figure 9. Measured values of viscosity η of tangerine and lemon juices as a function of concentration along the various selected isotherms (Magerramov et al. [15]). ×, Alvarado and Romero [92].

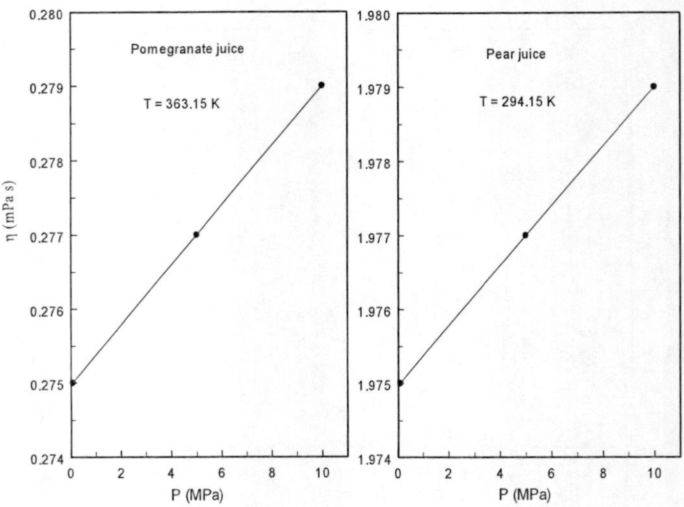

Figure 10. Viscosity of pomegranate and pear juices as a function of pressure along two selected isotherms 294.15 and 363.15 K (Magerramov et al. [14]).

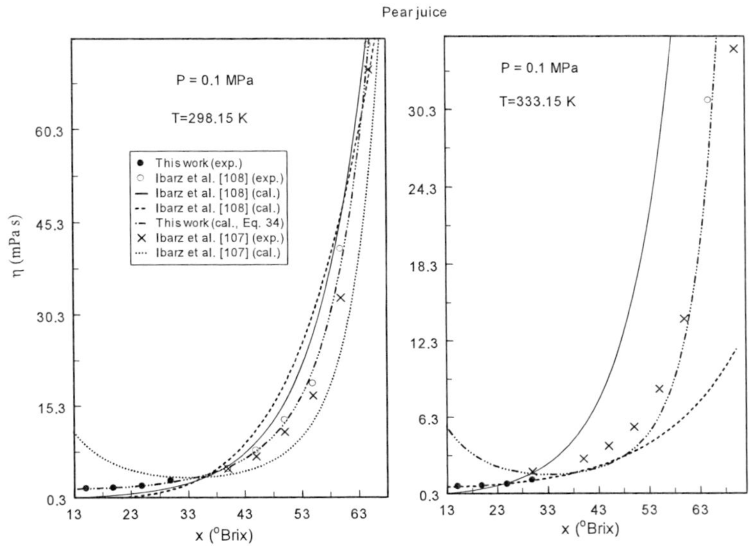

Figure 11. Comparisons of the concentration dependence of the measured and calculated values of the viscosities of pear juice using various models at fixed temperatures of 298.15 and 333.15 K and at atmospheric pressure.

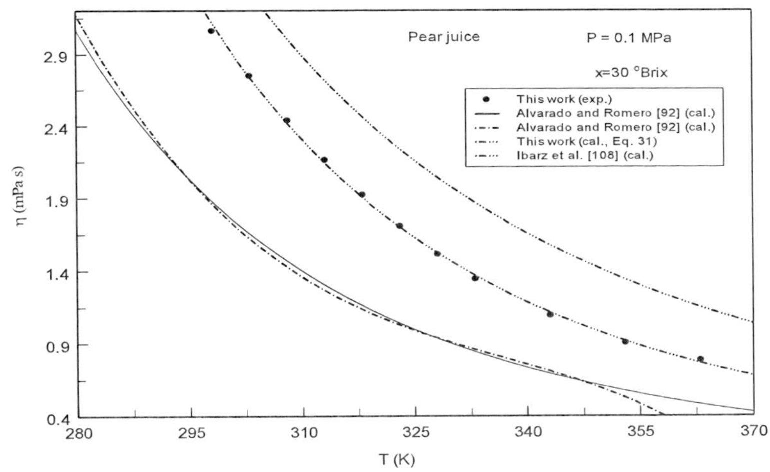

Figure 12. Comparisons of the temperature dependence of the measured and calculated values of the viscosities of pear juice using various models at fixed concentration of 30 °Brix and at atmospheric pressure.

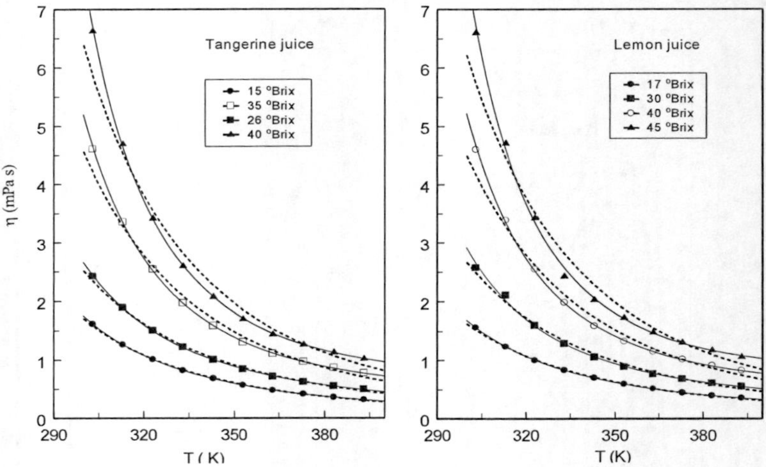

Figure 13. Comparisons of the temperature dependence of the measured (Magerramov et al. [15]) and calculated values of the viscosities of tangerine and lemon juices using various models at fixed concentrations. (------), calculated with original Arrhenius Eq. (31); (———), calculated with extended Arrhenius Eq. (32).

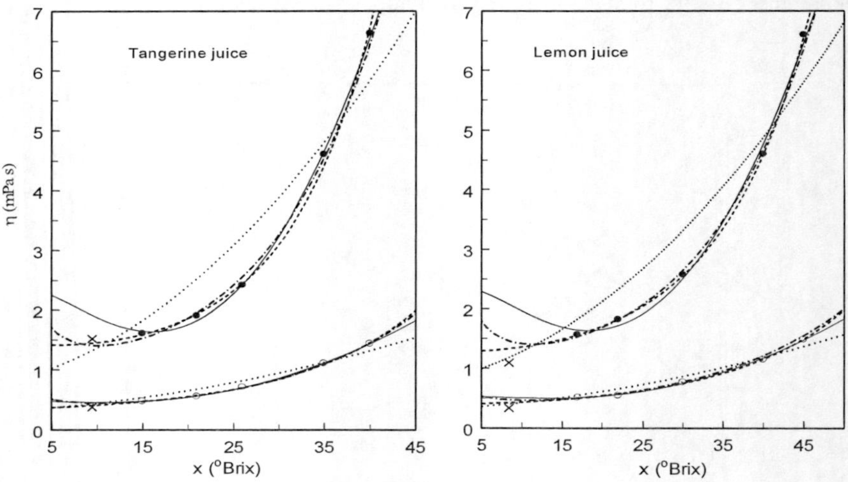

Figure 14. Comparisons of the concentration dependence of the measured and calculated values of the viscosities of tangerine and lemon juices by using the various models at two selected temperatures (303.15 and 363.15) K. •, this work (303.15 K); ○, this work (363.15 K); ×, Alvarado and Romero [92]. (———), Eq. (37); (--------), Eq. (34); (-· - · · - -), Eq. (35); (·········), Eq. (40).

The comparison of the datasets given in Table 2 for various fruit juices was difficult because the viscosity and other thermophysical properties of fruit juices are depend on various factors such as composition and soluble solids content due to: fruit type, generic characteristics, variety, ripening, place in the plant, size, plant nutritive level, agricultural practices, weather. This could explain the published data discrepancy (see below) for fruit juices. Below we briefly discussed the various datasets for some fruit juices.

Apple Juice

Nine datasets [38,93,98,107,110,112,159,160,161] were found in the literature for apple juice. Different techniques (capillary tube, coaxial cylinder, rotational viscometers) were used to measure the viscosity of apple juices. These data cover the temperature range from -5 to 80 °C and concentrations from 11 to 75 °Brix. Figure 15 demonstrate the comparison between the experimental data by Ibraz et al. [107] and the data reported by Alvarado and Romero [92] and the values calculated with the model by Bayindirli [93] and Rao [110]. At low concentrations (up to 40-45 °Brix) the agreement between the various datasets is satisfactory (dscrapances within 5-10 %), while at high concentrations (above 50°Brix) the deviations are very large. Probably this is due to difference in physical and chemical characteristics of the juices. Unfortunately some authors is not provide detailed information about physical and chemical characteristics of the studied sample of fruit juices. This is making difficult to estimate the reliability of the various reported datasets for the viscosity of fruit juices.

Orange Juice

Seventeen datasets [2,39,101,109,113,162-173] were reported by various authors in the literature. These data cover the temperature range from -19 to 80 °C and the concentration from 7 to 69.5 °Brix. Majority of the reported data were derived by using the coaxial cylinders and capillary viscometers. The direct comparison between the various reported viscosity data for orange juice and the values calculated with various viscosity models at selected isotherm of 303.15 K is presented in Figure 16. As one can see, the consistence between various datasets is satisfactory. Good agreement is found at low concentrations, while at high concentration the scattering various datasets is large.

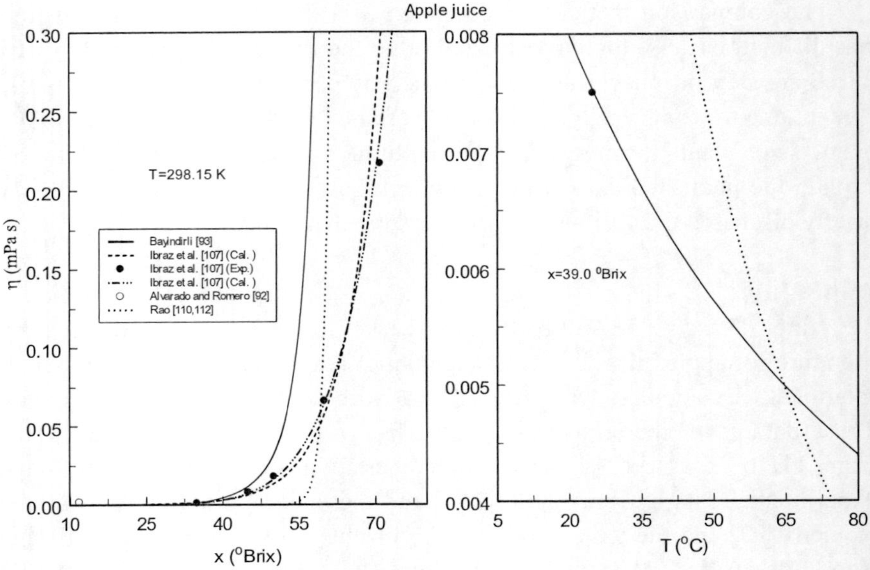

Figure 15. Experimental and calculated values of viscosity of apple juices as a function of concentration (left) and temperature (right) reported by various authors.

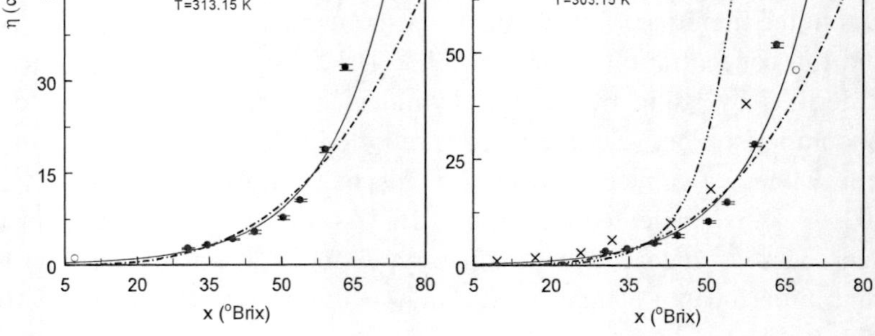

Figure 16. Measured and calculated with various models values of the viscosity of orange juices at two selected temperatures.

Grape Juice

Six datasets are available in the literature for the grape juice [38,110,112, 160,174,175]. Three different techniques, capillary tube, coaxial cylinders, and rotational viscometer, were used to perform the measurements. Measurements were made over the wide temperature range from -5 to 80 °C at concentrations between 0 and 75 °Brix. Figure 17 provide the comparison between the values of the viscosity of grape juice calculated with the models reported by Bayindirli [95] and Constenla et al. [98] at selected isotherm of 293.15 K and fixed concentration of 13.9 °Brix. The agreement is only qualitatively. The discrepancy is probably due to difference in the chemical composition of the grape juice sample.

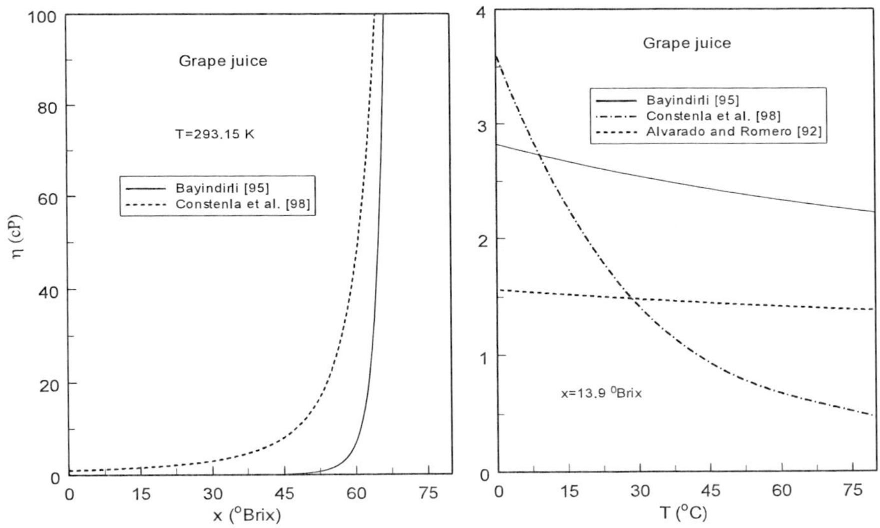

Figure 17. Viscosity of grape juice as a function of concentration (left) and temperature (right) calculated with various models reported in the literature.

Cherry Juice

Only two datasets are available for cherry juice Juszczak and Fortuna [100] and Giner et al. [183]. Both datasets were derived by using the coaxial cylinders technique. The data cover the range from 5 to 70 °C and concentrations from 22 to 74 °Brix. The uncertainty of the reported viscosity data by Genier et al. [183] vary within 0.08 to 190 mPa·s (or 2.77 to 5.9 %) depending on temperature and concentration. Figure 18 demonstrate the comparison between the both datasets for cherry juice in the $\eta - T$ and $\eta - x$ projections for the selected isotherms and concentrations. As one can see, the agreement is acceptable because the uncertainty of the values of viscosity calculated with Juszczak and Fortuna [100] model within 11-12 %. No physical and chemical characteristics of juice are presented in the work by Juszczak and Fortuna [100]. Both authors used the same viscosity model (see below section 4.4 Eqs. 44 and 47) to represent own experimental viscosity data. As one can see from Figure 18, these models fails to represent the experimental data at low temperatures and high concentrations.

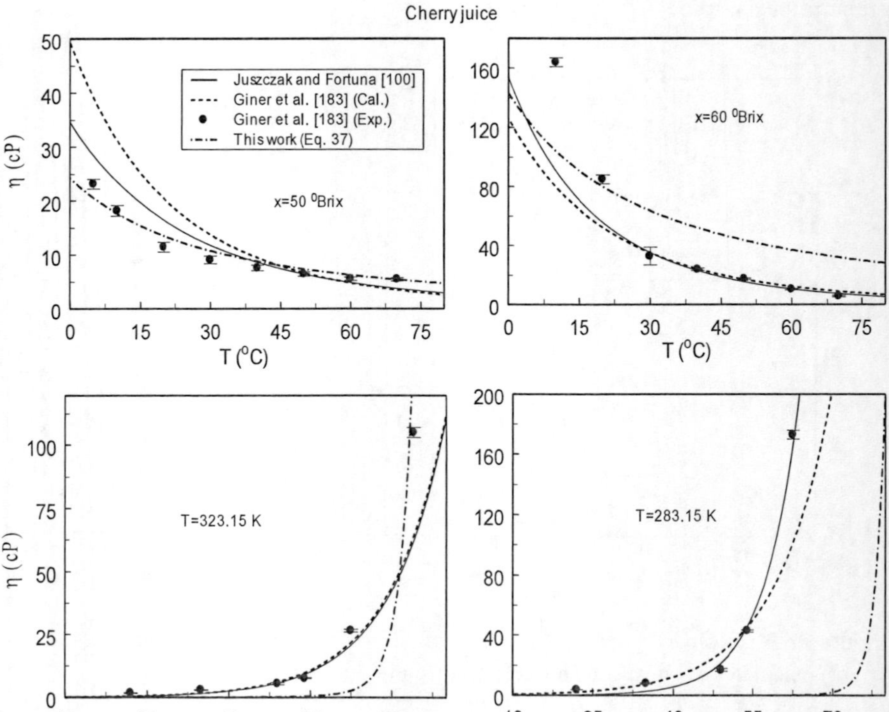

Figure 18. Experimental and calculated values of the viscosity of cherry juice as a function of temperature and concentration reported by various author in the literature.

Pear Juice

Four datasets were found for pear juice [15,92,107,108]. Two different (coaxial cylinders and capillary tube) techniques were employed to measure the viscosity of pear juice at temperatures from 5 to 130 °C and at concentrations between 8 and 71°Brix. The measurements by Magerramov et al. [15] were performed under pressure. The detailed comparison various data sources and prediction model for pear juices is presented in section 4 (see Figures 11 and 12).

Watermelon

Two datasets were reported in the literature for watermelon juice by Alvarado and Romero [92] and Sogi [194]. Measurements by Alvarado and Romero [92] were performed as a function of temperature along the only one concentration of 7.2 °Brix, while Sogi [194] made the measurements at temperatures from 5 to 50 °C and at concentrations from 23 to 42 °Brix. These data were derived by using the capillary tube [92] and rotational [194] viscometers. The comparison between the models reported by Sogi [237] and Alveredo and Romero [92] for watermelon is presented in Figure 19. As one can see the agreement is acceptable.

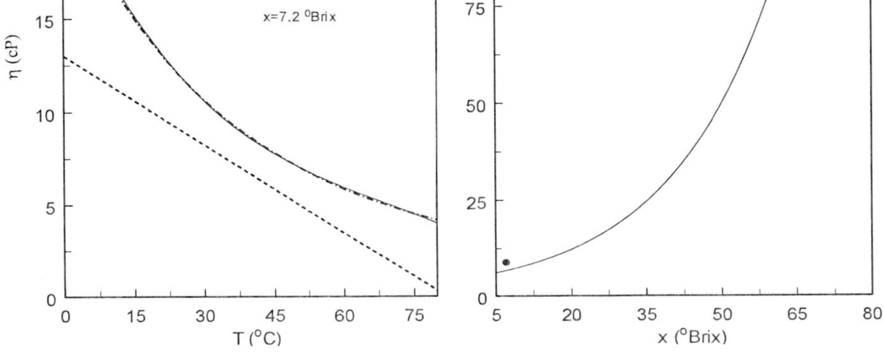

Figure 19. Temperature and concentration dependences of the viscosity of watermelon juice calculated with various models.

Pineapple Juice

Viscosity data for this juice reported by Alvarado and Romero [92], Garciá et al. [187], and Varshney and Kumbhar [168]. The reported data are cover the temperature range from 22 to 45 °C and concentrations range was between 7 and 36 °Brix. Measurements were performed with coaxial cylinder [168] and capillary viscometers [92,187]. The viscosity data for pineapple juices by Garciá et al. [187] and Varshney and Kumbhar [168] together with values calculated from correlation by Alvarado and Romero [92] at selected concentration of 13.9 °Brix are presented in Figure 20. The comparison of the viscosity results reported by various authors is difficult because of temperature and concentration differences between the measurements. Therefore, some time to comparison the various datasets reported by other authors in the literature, we used an interpolating procedure (analytical method). As Figure 20 shows, the agreement between various datasets and prediction models is reasonable.

Pomegranate Juice

The viscosity of pomegranate juice was measured by several authors [15,114,184,185]. The measured range of temperature is from 5 to 130 °C. The concentration ranged from 11 to 75 °Brix. The capillary tube and stress rheometer were used to measure the viscosity. The lack of the enough information provided by the authors in the publications is precluded the comparison reported viscosity data for pomegranate juices. Figures 6,8, and 10 provide the experimental temperature, concentration, and pressure dependences of the viscosity of pomegranate juices together with values calculated from Eq. (32).

Mango Juice

Three datasets were reported in the literature by Manohar et al. [198], Rao et al. [111], and Singh and Eipeson [199] for mango juices. Measurements were made with coaxial cylinder and rotational viscometers in the temperature range from 15 to 85 °C and at concentrations between 11 and 66 °Brix. The measured values of viscosity for mango juices were used by Singh and Eipeson [199] to calculate the Arrhenius equation parameters as a function of concentration in the range from 15 to 66 °Brix. The concentration dependence of the Arrhenius equation parameters was presented as exponential function.

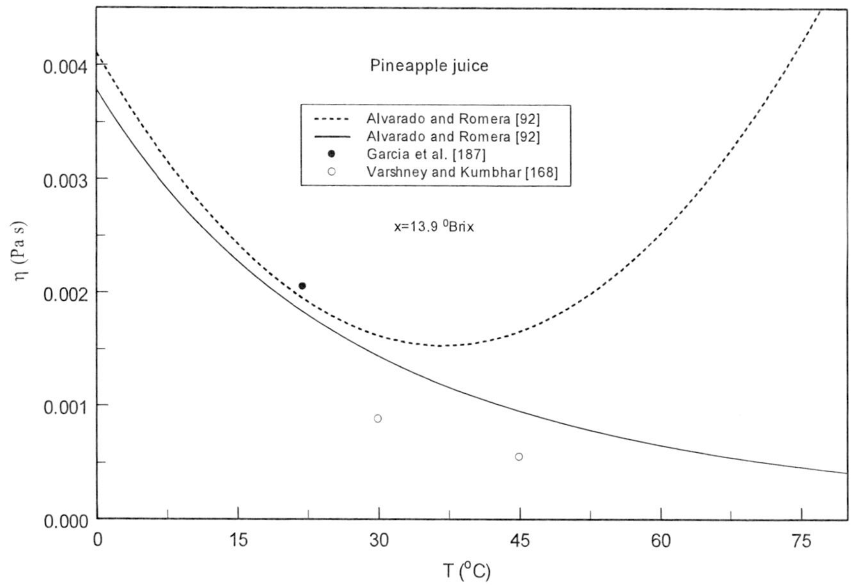

Figure 20. Effect of temperature on the viscosity of pineapple juice at selected concentration.

Lemon and Tangerine Juices

Only two datasets were found for lemon and tangerine juices (Alvarado and Romero [92] and Magarremov et al. [15]). These data cover the range of temperature from 10 to 119 °C and concentrations between 8.4 and 45 °Brix. Measurements were made with capillary viscometer. The measurements by Magarremov et al. [15] were performed under pressure (up to 10 MPa). Alvarado and Romero [92] reported temperature dependence of the viscosity for lemon and tangerine juices only at one concentration of 8.4 and 9.4 °Brix, respectively. The details of the comparison various datasets and prediction models for lemon and tangerine juices is presented in section 4 (Figures 7 and 14).

Strawberry Juice

Two datasets were found also for strawberry juices (Juszczak and Fortuna [99] and Simpson [195]). Measurements were performed using the capillary tube and coaxial cylinders techniques in the temperature range from 10 to 60 °C and at concentration from 50 to 67.1 °Brix. The measured values of the viscosity by Juszczak and Fortuna [99] were used to calculate the parameters of the combined models (see section 4.4. Eqs. 44 and 47).

Guava Juice

Three datasets were found in the literature for this guava juices reported by Vitali and Rao [196], Garciá et al. [187], and Brekke and Myers [197]. All data were derived by using the capillary tube viscometer. The temperature and concentration ranges were between 22 and 60 °C and 7.2 to 40.8 °Brix, respectively.

Banana Juice

The viscosity of banana juices was measured by Garciá et al. [187] and Khalil et al. [175]. These data cover the temperature range from 22 to 70 °C and concentrations between 13 and 80 °Brix. Measurements were made with coaxial cylinder and capillary viscometers.

Tomato Juice

Seven datasets were found for this juice [176-182]. These data cover the temperature range from 10 to 95 °C and concentrations from 5 to 32 °Brix. Three different techniques (capillary tube, coaxial cylinder, and rotational viscometer) were used to measure the viscosity of the juice.

Cranberry, Raspberry, Peach, Blackcurrant, Celery, Mulberry, and Apricot, Papaya, Grapefruit Juices

For these juices only one reported viscosity dataset was found in the literature. These data were derived by using the coaxial cylinder [104-106,185,200], capillary tube [187], and rotational [167,194]. The comparison between the reported data is not possible due to lack the enough information provided by the authors.

Chapter 4

MODELING, PREDICTION AND CORRELATION TECHNIQUES

Various models were used in the literature to describe the temperature, concentration, and pressure dependences and their combined effect on the viscosity of liquid foods and aqueous solutions. Most frequently used by various authors models for the viscosity liquid foods and aqueous solutions are summarized in Table 4. Below we briefly reviewed the most often used models to describe the effect of temperature and concentration and their combined effect on the viscosity of fluids and fluid mixtures. We examined the accuracy of the models and their applicability for the fruit juices.

Table 4. Summary of the models used for the correlation of the viscosity of fruit juices and aqueous solutions.

Functional Form of the Models	Reference
Temperature dependency models at x=constant	
$\eta(T) = \eta_{\infty} \exp(E_a / RT)$	Juszczak and Fortuna [99,100], Ibraz et al. [102-107], Kaya and Sözer [114], Altan and Maskan [184], Khalil et al. [175], Rao and Anantheswaran [40]
$\log \eta = a + \dfrac{b}{T - c}$	Rao [19], Soesanto and Williams, [203]

Table 4. (Continued).

Functional Form of the Models	Reference
Temperature dependency models at x=constant	
$\eta = \eta_0 \exp\left(\frac{b}{T} + \frac{c}{T^2}\right)$	Magerramov et al. [14,15]
$log\,\eta = a_0 + a_1 T + a_2 T^2$	Rao [204]
$\eta = a_0 + a_1 T + a_2 T^2 + a_3 T^3$	Alvarado and Romero [92]
Concentration dependency models at T=constant	
$\eta(T) = \eta_1 x^n$, $\eta(T) = \eta_1\, exp(bx)$	Ibraz et al. [102-107], Juszczak and Fortuna [99, 100], Sogi [194], Altan and Maskan [184], Khalil et al. [175], Singh and Eipeson [199]
$\eta = \eta_3 a_W{}^n$, $\eta(T) = \eta_4\, exp(ba_W)$, a_W - water activity	Ibraz et al. [102-107]
$log\,\eta = 1.1x/(100 - 0.856x)$	Mizrahi [205]
$\eta_S = \eta_{wis} \exp[f(x)]$, $f(m) = a_1 x + a_2 x^2 + \cdots$	Leyendekkers and Hunter [129,130], Leyendekkers [131], Ibarz et al. [108]
$\eta = \frac{\eta_0 \exp(xE)}{1 + xV}$,	Goldsack and Franchetto [132,133], Chirife [128] Chirife and Buera [134]
$\eta/\eta_0 = 1 + A\sqrt{x} + Bx$, $\eta/\eta_0 = 1 + A\sqrt{x} + Bx + Dx^2$, $\eta/\eta_0 = 1 + A\sqrt{x} + Bx + Dx^2 + Fx^{2.5}$	Jones and Talley [206], Kaminsky [144-147], Desnoyers and Perron [149], Desnoyers et al. [148], Feakins and Lawrence [150], Robertson and Tyrrell [151], Abdulagatov et al. [127,141-143,152], Abdulagatov and Azizov [140,122-126]

Functional Form of the Models	Reference
Concentration dependency models at T=constant	
$\eta / \eta_0 = 1 + d_1 x + d_2 x^2 + d_3 x^3$	Kestin and Shankland [61,75], Grimes et al. [74]
$\eta / \eta_0 = 1 + 2.5\phi + 10.05\phi^2 = 1 + 2.5V_k x + 10.$	Thomas [153], Charm [207]
$\eta / \eta_0 = 1 + 2.5\overline{V}x + k_1 \overline{V}^2 x^2 + k_2 \overline{V}^3 x^3 + k_3$	Sahu and Behera [157]
Temperature and concentration dependency models at P=constant (Combined models for the viscosity of fruit juices)	
$\eta(T) = \eta_3 \exp\left(E_a / RT + bx\right)$, $\eta(T) = \eta_4 x^n \exp\left(E_a / RT\right)$	Juszczak and Fortuna [99,100], Alvarado and Romero [92], Altan and Maskan [184], Khalil et al. [175]
$\eta(T) = \eta_3 \exp\left(E_a / RT + b a_W\right)$, $\eta(T) = \eta_3 \left(a_W\right)^n \exp\left(E_a / RT\right)$	Ibarz et al. [102-107]
$\eta = a_0 \left(9/5t + 32\right)^{0.9982} \left(b_0 + b_1 x + b_2 x^2\right)$	Balint [208]
$\eta = \exp\left(a + b/T + cx\right)$	Altan et al. [158]
$\eta = \eta_0 \exp\left(\dfrac{(a + b/T)x}{100 - (c + dT)x}\right)$	Bayindirli [93-96,209]
$\eta = K_2 \exp\left(\dfrac{E_a}{RT} + K_3 x + K_4 x^2\right)$	Ibraz et al. [108]
$\eta(T) = \eta_0 \exp\left(b/T\right)$, $\ln \eta_0 = \sum_{i=0}^{4} a_i x^i$, $b = \sum_{i=0}^{4} b_i x^i$	Magerramov et al. [14,15]
$\eta / \eta_0 = 1 + A\sqrt{x} + Bx + Dx^2$, $A(t) = \sum_{i=0}^{3} A_i t^i$, $B(t) = \sum_{i=0}^{3} B_i t^i$, $D(t) = \sum_{i=0}^{3} D_i t^i$	Magerramov et al. [14,15]

Table 4. (Continued).

Functional Form of the Models	Reference
Temperature and concentration dependency models at *P*=constant (Combined models for the viscosity of fruit juices)	
$\log\eta = a_0 + a_1T + a_2T^2 + (b_0 + b_1T + b_2T^2)x + (c_0 + c_1T + c$	Fernéndez-Martin [210]
$\log_{10}\eta = a_0N - a_1(30 - T)(a_2 + a_3N^{1.25})/(91$, $N = x(1900 - 18x)$	Peacock [211]
$\eta/\eta_0 = 1 + d_1x + d_2x^2 + d_3x^3$, $d_i = \sum_{j=0}^{3} d_{ij}t^j$	Kestin and Shankland [61,75], Grimes et al. [74]
Models which represent the viscosity of juice relative to pure water viscosity (see also Table 11)	
$\eta_r = 1 + ax/(1 + bx^{1/2}) + dx$, $\eta_r = \eta/\eta_0$ is the relative viscosity	Elfak et al. [212]

*
$$\log\frac{\eta_0(\theta)}{\eta(20\ ^\circ\mathrm{C})} = \frac{\theta}{116-\theta}\{1.2378 - 1.303\times10^{-3}\theta + 3.06\times10^{-6}\theta^2 + 2.55\times10^{-8}\theta^3\},$$

Where θ=20– T/°C, is the viscosity of pure water; The estimated uncertainty of equation is better than ±0.1% (Kestin et al. [73])

4.1. Temperature Dependence of the Viscosity of Fruit Juices

The experimental temperature dependences of the viscosity of fruit (pomegranate, pear, tangerine, and lemon) juices along the selected constant concentrations are presented in the Figures 6 and 7 as an example of the typical temperature effect on the viscosity of fruit juices. The viscosity of fruit juices considerably decreases with temperature. For example, as one can see from Figure 6 (left) (see also Table 1 in Appendix), at constant pressures (from 0.1 to 10 MPa) between temperatures (293 and 403) K the viscosity of pomegranate juice is significantly (by a factor of 4 to 7) affected by temperature at high concentrations (above 30 °Brix). However, at low concentrations (below 23 °Brix), the temperature is little (by a factor of 1 to 2) influences viscosity. The effect of temperature on viscosity of fruit juice is different for distinct temperature range. The viscosity of juices is considerably affected by temperature below 360 K. At high temperatures (above 360 K) the rate of the viscosity decreasing with temperature is small. For example, at constant concentrations (from 15 to 40 °Brix) between temperatures (303 and 360) K the viscosity of tangerine and lemon juices changes by factor of 4.5, while at high temperatures (above 360 K) the viscosity changes by factor of 1.5 only. The temperature effect on viscosity strongly depends also on the concentration of juice. At low concentrations (below 30 °Brix) the rate of viscosity changes is small (changes by factor 4), while at high concentrations the rapid changes (by factor 6) of viscosity with temperature is observed (see Figures 6 and 7). To quantitatively estimate the effect of temperature on the viscosity of juices the temperature coefficient of viscosity, $(\partial \ln \eta / \partial T)_x$ was calculated from the measured values of viscosity for tangerine and lemon juices. The temperature coefficient of viscosity for tangerine juice is changes within (0.007 to 0.04) K^{-1} depending on concentration and temperature. At low temperatures the rate of temperature changes of the viscosity, $(\partial \eta / \partial T)_x$, of tangerine juice changes within (0.025 to 0.040) K^{-1}. The highest rate up to 0.04 K^{-1} was found in the low temperature range and high concentrations.

Table 4 provides the list of the various models which were used in the literature to describe the temperature dependence of the viscosity of liquid foods and aqueous solutions. The empirical equation of Arrhenius is valid for temperature dependence of the viscosity of liquid foods [14,15,18,19,29,33,40,92-98,99-108,110,111,112,113-115,116,117]

$$\eta = \eta_0 \exp\left(\frac{b}{T}\right), \tag{31}$$

where η_0 (values of viscosity at high temperatures, $T \to \infty$) and $b = E_a / R$ (E_a is the flow activation energy) are function of concentration. This relation also often used to represent experimental viscosity data for pure fluid and fluid mixtures (for example, aqueous solutions, see [118-127]). However, Eq. (31) failed to represent experimental viscosity data at high temperatures and high concentrations. One more term is required to accurate describe the experimental viscosity data at high temperatures

$$\eta = \eta_0 \exp\left(\frac{b}{T} + \frac{c}{T^2}\right). \tag{32}$$

Equation (31) was theoretically rigorous substantiated by the Eyring's absolute rate theory for concentrated aqueous electrolyte solutions (Glasstone et al. [118]) in the form

$$\eta = \frac{hN}{V}\exp\left(\frac{\Delta G^+}{RT}\right) \text{ or } \eta = A\exp\left(\frac{\Delta H^+}{RT}\right), \tag{33}$$

where ΔG^+ and ΔH^+ are the free enthalpy of activation and enthalpy of activation, h is Planck's constant, N Avogadro's constant, and R the gas constant, V is the molar volume.

The measured values of the viscosity of pomegranate, pear, tangerine, and lemon juices were expressed by the Arrhenius relationship (31) and by Eq. (32). The values of the Arrhenius parameters for pomegranate, pear, tangerine, and lemon juices calculated with the measured viscosity data at atmospheric pressure and at pressures of 5 and 10 MPa as a function of concentration are given in Table 5. This equation represents experimental values of viscosity for fruit juices within 0.65 % in the temperature range from 292 to 363 K. The experimental viscosity data for pear juice as a function of temperature are shown in Figure 12 at selected concentration of 30 °Brix together with the values calculated from Eq. (31) and various correlation models reported by other authors [92,108].

As one can see from Table 5, the values of activation energy for the flow E_a are increases with concentration of juices. The flow activation energy E_a and parameter $\ln\eta_0$ can be directly calculated from the slope and intersect of the straight line by the Arrhenius relationship function ($\ln\eta \approx 1/T$) (Arrhenius plot $\ln\eta$ vs. $1/T$, see Figure 21).

Table 5. The Arrhenius Eqs. (31) and (32) parameters ($\ln \eta_0$, $b = E_a / R$, and c) for tangerine and lemon juices as a function of concentration.

Equation (31)

Tangerine juice			Lemon juice		
x (°Brix)	$\ln \eta_0$ (mPa s)	$b = E_a / R$ (K)	x (°Brix)	$\ln \eta_0$ (mPa s)	$b = E_a / R$ (K)
15	-6.68899	2165.783	17	-6.10088	1975.496
21	-6.51205	2159.819	22	-6.35765	2097.062
26	-6.17033	2127.509	30	-5.98611	2089.936
35	-6.36895	2365.660	40	-6.09457	2279.163
40	-6.39998	2476.157	45	-6.15909	2395.723
Pomegranate juice (0.1 MPa)			Pear juice (0.1 MPa)		
x (°Brix)	$\ln \eta_0$ (mPa s)	$b = E_a / R$ (K)	x (°Brix)	$\ln \eta_0$ (mPa s)	$b = E_a / R$ (K)
11	0.8590	1898.44	15.2	0.7299	2014.76
23	0.5901	2049.47	20.0	0.7238	2037.26
30	-0.0457	2382.97	25.0	0.6444	2106.70
35	-0.5544	2650.71	30.0	0.1817	2343.95
40	-0.4099	2711.51	40.0	-0.4099	2711.51
Pomegranate juice (5 MPa)			Pear juice (5 MPa)		
11.0	1.0956	1817.28	15.2	0.9075	1958.77
Pomegranate juice (10 MPa)			Pear juice (10 MPa)		
11.0	0.9505	1871.97	15.2	1.3376	1826.50

Table 5. (Continued).

Equation (32)

Tangerine juice				Lemon juice			
x (°Brix)	$\ln \eta_0$	b	$c \times 10^{-5}$	x (°Brix)	$\ln \eta_0$	b	$c \times 10^{-5}$
15	-4.5578	693.639	2.5247863	17	-3.4798	164.955	3.1051494
21	-3.1557	-158.638	3.9762453	22	-2.0487	-879.422	5.1047882
26	-2.0514	-717.649	4.8795601	30	0.5156	-2401.194	7.7024672
35	3.2081	-4249.82	11.345803	40	4.8281	-5265.764	12.939851
40	6.2801	-6282.718	15.021821	45	8.6979	-7866.905	17.600817

The intercepts and slopes of the linear plots ($\ln\eta \approx 1/T$) are flow activation energy E_a and parameter $\ln\eta_0$, respectively. As one can see from Figure 21, $\ln\eta$ *vs.* $1/T$ curves are straight line only for the low concentrations (see also Erickson et al. [219]).

Therefore, the curvature of the $\ln\eta$ *vs.* $1/T$ curves at high concentration can be taking into account by relation (32). Figure 13 compare the present results for viscosity with values calculated from Eqs. (31) and (32). As one can see, at low concentrations (below 26 °Brix) both models good represent the experimental data. However, at high concentrations extended Arrhenius model (32) considerable accurate reproduced measured values of the viscosity of tangerine and lemon juices than original Arrhenius model (31). Therefore, Eq. (32) can be recommended for future used to represent experimental viscosity data for fruit juices at high temperatures and high concentrations.

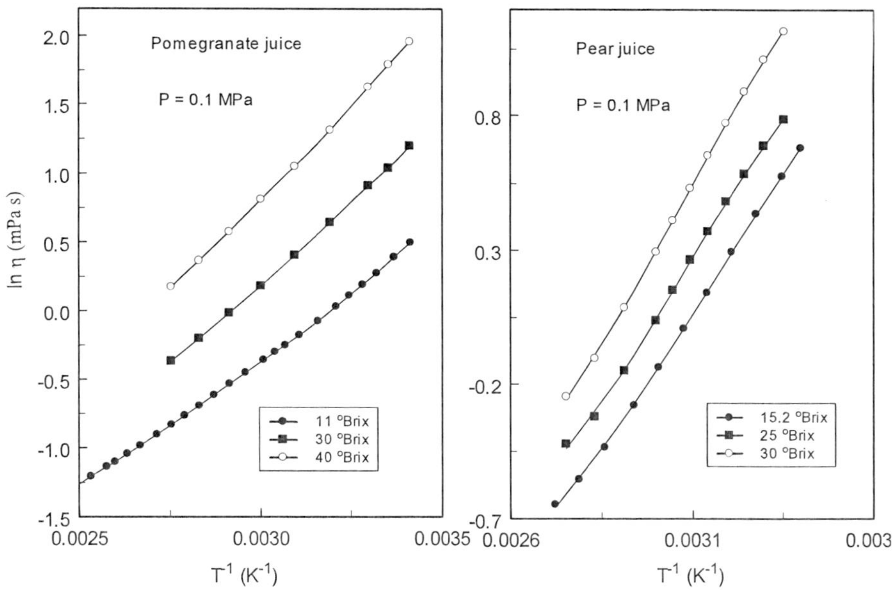

Figure 21. Experimental $ln\,\eta$ as a function of T^{-1} (Arrhenius plot) for pomegranate and pear juices (Magerramov et al. [14]).

4.2. Concentration Dependence of the Viscosity of Fruit Juices

Figures 8 and 9 show the concentration dependence of the viscosity of fruit juices along various selected isotherms at atmospheric pressure. Figure 8 demonstrate the effect of concentration on the viscosity of pomegranate and pear juices at various fixed temperatures. As one can see from Figure 8, the viscosity of pomegranate and pear fruit juices observed considerably increases (up to 2.2 % to 3.5 %) at concentration above 25 °Brix, especially at low temperatures. The viscosity of tangerine and lemon juices (see Figure 9) in the concentration range from (15 to 45) °Brix changes by factor 3.2 to 4.2. These figures also demonstrate how the behavior of the concentration dependence of the viscosity of fruit juices depends on temperature. As Figures 8 and 9 shows, the viscosity of fruit juices monotonically increases with the concentration. However, the effect of concentration on the viscosity of juices strongly depends on temperature and concentration. For example, at concentrations x<20 °Brix the viscosity is little affected (by factor 1.2) by concentration, while at high concentrations the effect of concentration stronger (by factor 3.5). The rate of the viscosity changes with concentration is depends on temperature also. At low temperatures (below 343 K) the slopes of the $\eta - x$ curves are small (changes by factor 3), while at high temperatures (above 343 K) the rapid changes (by factor 4) of the viscosity with concentration are observed. The concentration coefficient of viscosity, $(\partial \ln \eta / \partial x)_T$, for the juices was calculated using the measured values of the viscosity of fruit juices. The values of concentration coefficient of viscosity for tangerine juice is changes within (0.3 to 4.5) °Brix^{-1} depending on concentration and temperature. The highest rate, $(\partial \ln \eta / \partial x)_T$, up to 4.5 °Brix^{-1} was found in the low temperature range and high concentrations.

There are different theoretical models to represent the concentration dependence of liquid foods and aqueous solutions (see Table 4). Below we briefly discussed the most often used models to describe the effect of concentration on the viscosity of fluids and fluid mixtures and their applicability for fruit juices.

Leyendekkers and Hunter [129,130] and Leyendekkers [131] have applied the TTM (Tammann-Tait-Gibson) model to the calculation of viscosity of the aqueous electrolyte solutions. According TTG model the viscosity equation can be present as

$$\eta_S = \eta_{wis} \exp[f(x)],\ f(x) = a_1 x + a_2 x^2, \tag{34}$$

where η_S and η_{wis} represent the viscosities of the solution (or juice) and the water in solution (or in juice), respectively, and x is the concentration. This relation can be used also to represent the concentration dependence of the viscosity of juices. Figures 11 and 14 depicts the values of the viscosity of pear, tangerine, and lemon juices calculated with Eq. (34) together with other models. The values of parameters of Eq. (34) for some fruit juices derived from the present measurements are given in Table 6. Ibarz et al. [108] also used this relation to accurate represent measured values of the viscosity for pear juice. The agreement between measured and calculated values of viscosity is excellent (AAD is within 0.5 to 1.0 %).

Goldsack and Franchetto [132,133] proposed the simplified form of the Eq. (33) to describe the concentration dependence of the aqueous solutions

$$\eta = \frac{\eta_0 \exp(xE)}{1 + xV}, \tag{35}$$

where η is the viscosity of the solution or juice at a concentration x and temperature T, η_0 is the viscosity of the solvent (pure water) at temperature T, x is the concentration, and E and V are the adjustable parameters. The temperature dependence of the viscosity of fruit juices can be explained in terms of the temperature dependence of the E and V parameters of an Eq. (35). Unfortunately, the theory cannot to predict the temperature dependence of the E and V. Therefore, the values of the parameters E and V can be extracted from the experimental data as a function of temperature by fitting each isotherm to the Eq. (35). The results of the application of Eq. (35) to the measured viscosities of pomegranate, pear, tangerine, and lemon juices are presented in Table 7. This model has been used by Chirife [128] and Chirife and Buera [134] to fit viscosity data of various sugar and sugar mixtures up to very high concentrations.

The existing theoretical result which describes the concentration dependence of the viscosity of aqueous solutions is valid only at very dilution.

Table 6. The values of parameters $\ln\eta_{wis}$ and a_i (Eqs. 34) for fruit juices as a function of temperature.

Tangerine juice				Lemon juice		
T (K)	$\ln\eta_{wis}$ (mPa s)	$a_1 \times 103$ (°Brix-1)	$a_2 \times 103$ (°Brix-2)	$\ln\eta_{wis}$ (mPa s)	$a_1 \times 103$ (°Brix-1)	$a_2 \times 103$ (°Brix-2)
303.15	0.38913	-14.9364	1.32523	0.24826	-3.5141	0.88719
313.15	0.09069	-8.2186	1.12571	-0.20719	14.8562	0.53334
323.15	-0.20402	-0.4041	0.91634	-0.31373	7.8867	0.58774
333.15	-0.46844	5.5330	0.75972	-0.61370	18.2529	0.34286
343.15	-0.67651	8.0157	0.68460	-0.66400	8.7517	0.48471
353.15	-0.86352	9.0915	0.64927	-0.74775	2.2142	0.59102
363.15	-1.06047	11.1707	0.61624	-0.93747	4.8031	0.55168
373.15	-1.21220	11.4476	0.62303	-1.12172	8.1858	0.50707
383.15	-1.40096	15.1601	0.57764	-1.28241	10.9391	0.46989
393.15	-1.54684	16.7374	0.56530	-1.43307	13.6984	0.43332
Pomegranate juice				Pear juice		
T (K)	$\ln\eta_{wis}$ (mPa s)	a_2 (°Brix-1)	a_2 (°Brix-2)	$\ln\eta_{wis}$ (mPa s)	a_1 (°Brix-1)	a_2 (°Brix-2)
293.15	0.8609	-5.8990	2.303	-	-	-
298.15	0.7103	-5.6930	2.237	1.2596	-8.5280	2.6796
313.15	0.2447	-4.2933	1.838	0.7090	-6.1100	2.1005
333.15	-0.1575	-3.7152	1.587	0.3459	-6.2035	2.0054
Pomegranate juice				Pear juice		
T (K)	$\ln\eta_{wis}$ (mPa s)	a_1 (°Brix-1)	a_2 (°Brix-2)	$\ln\eta_{wis}$ (mPa s)	a_2 (°Brix-1)	a_2 (°Brix-2)
353.15	-0.5411	-2.8335	1.302	-0.0667	-5.3504	1.7390
363.15	-0.7754	-1.6360	0.993	-0.5167	-2.0361	0.9744

Table 7. The values of parameters V and E (Eq. 35) for fruit juices as a function of temperature.

Tangerine juice				Lemon juice			
T (K)	η_0 (mPa s)	E (°Brix^{-1})	V (°Brix^{-1})	T (K)	η_0 (mPa s)	E (°Brix^{-1})	V (°Brix^{-1})
303.15	24.5624	0.09662	4.50	303.15	18.6009	0.08545	3.00
333.15	18.8733	0.08579	5.50	333.15	0.93089	0.09209	0.15
363.15	1.17754	0.08080	0.49	363.15	0.95924	0.07136	0.32
Pomegranate juice				Pear juice			
T (K)	η_0 (mPa s)	E (°Brix^{-1})	V (°Brix^{-1})	T (K)	η_0 (mPa s)	E (°Brix^{-1})	V (°Brix^{-1})
293.15	324.27	0.1022	67.988	298.15	180.890	0.0843	25.322
303.15	128.25	0.0970	30.662	313.15	161.211	0.0813	28.984
313.15	85.799	0.0930	23.984	323.15	78.5376	0.0800	17.134
323.15	32.125	0.0910	10.723	333.15	42.5676	0.0760	10.492
363.15	16.011	0.0795	8.1656	343.15	36.0353	0.0750	10.532
-	-	-	-	363.15	9.4179	0.0690	3.1565

Falkenhagen – Onsager – Fuoss [135,136] and Debye-Hückel-Onsager [137,138] theory predicts a square root concentration, $\frac{\eta}{\eta_0} \propto \sqrt{x}$ (limiting law), dependence of the viscosity of solutions at infinite dilution ($x \to 0$).

This theory correctly explains the rise of viscosity with concentration in the limit of very low (dilute solutions) concentrations. The dilute solution theories [135-138] of viscosity are not practicable because of their very limited concentration range. Jones and Dole [139] proposed an empirical extension of the previous model [135] to high concentrations as

$$\frac{\eta}{\eta_0} = 1 + A\sqrt{x} + Bx \,. \tag{36}$$

In Eq. (36) η and η_0 are the viscosities of an electrolyte solution and pure solvent (water), respectively, A is an always positive constant, and x is the concentration. Several authors [140-143,133,144-151], included a quadratic term Dc^2 (extended Jones-Dole equation)

$$\frac{\eta}{\eta_0} = 1 + A\sqrt{x} + Bx + Dx^2 \,, \tag{37}$$

to extend the Jones-Dole equation for more concentrated solutions. Abdulagatov et al. [127, 141-143,152] added one more term $Fc^{2.5}$ for application Eq. (37) to higher concentration (up to saturation)

$$\frac{\eta}{\eta_0} = 1 + A\sqrt{x} + Bx + Dx^2 + Fx^{2.5} \,. \tag{38}$$

The measured relative viscosities for pomegranate, pear, tangerine and lemon juices were fitted with the Eqs. (37) (Magerramov et al. [14,15]). The derived values of the parameters are given in Table 8 as a function of temperature. Good agreement (within 0.5 to 1.2 %) was found between measured and calculated with Eq. (37) values of viscosity.

Kestin and Shankland [75] and Grimes et al. [74] omitted the square-root term owing to its small magnitude to describe the concentration dependence of the experimental viscosity data for aqueous solutions. The experimental viscosity data for aqueous solutions were fitted to

Table 8. The values of parameters *A*, *B*, and *D* (Eq. 37) for fruit juices as a function of temperature.

Tangerine juice				Lemon juice			
T (K)	A (°Brix-1/2)	B (°Brix-1)	D (°Brix-2)	T (K)	A (°Brix-1/2)	B (°Brix-1)	D (°Brix-2)
303.15	2.2595	-0.7144	0.013450	303.15	2.1389	-0.6362	0.010566
313.15	1.7283	-0.5398	0.010497	313.15	1.3524	-0.3983	0.007366
323.15	1.2532	-0.3866	0.007977	323.15	1.1645	-0.3372	0.006196
333.15	0.9540	-0.2918	0.006367	333.15	0.5575	-0.1524	0.003620
343.15	0.7997	-0.2442	0.005512	343.15	0.7314	-0.2042	0.004083
353.15	0.6815	-0.2105	0.004915	353.15	0.8112	-0.2308	0.004329
363.15	0.5931	-0.1877	0.004566	363.15	0.7101	-0.2045	0.004001
373.15	0.5859	-0.1889	0.004589	373.15	0.6514	-0.1900	0.003863
Pomegranate juice				Pear juice			
T (K)	A (°Brix-1/2)	B (°Brix-1)	D (°Brix-2)	T (K)	A (°Brix-1/2)	B (°Brix-1)	D (°Brix-2)
298.15	1.4779	-0.5033	0.01029	298.15	1.9165	-0.5829	0.01043
303.15	1.3383	-0.4545	0.00940	303.15	1.9028	-0.5755	0.01028
313.15	1.1312	-0.3799	0.00792	313.15	1.5249	-0.4516	0.00830
323.15	1.0153	-0.3383	0.00705	323.15	1.4000	-0.4132	0.00759
333.15	0.9378	-0.3086	0.00634	333.15	1.2335	-0.3594	0.00654
343.15	0.8223	0.2672	0.00551	343.15	1.1462	-0.3333	0.00601
353.15	0.7149	-0.2310	0.00482	353.15	1.0755	-0.3099	0.00549
363.15	0.5871	-0.1859	0.00400	363.15	0.6807	-0.1774	0.00342

$$\frac{\eta}{\eta_0} = 1 + d_1 x + d_2 x^2 + d_3 x^3 , \qquad (39)$$

where $d_i = \sum_{j=0}^{3} d_{ij} t^j$ are the function of temperature.

Thomas [153] has extended the Einstein relation (Einstein [154]) for the hydrodynamic effect to high concentrations by showing that for suspensions the relative viscosity is given by the relation

$$\frac{\eta}{\eta_0} = 1 + 2.5\phi + 10.05\phi^2 = 1 + 2.5V_k x + 10.05V_k^2 x^2 . \tag{40}$$

Table 9. The values of V_k (Eq. 40) for fruit juices as a function of temperature.

T, K	$V_k \times 103$	T, K	$V_k \times 103$
Tangerine juice		Pomegranata juice	
303.15	0.3639	293.15	0.2805
313.15	0.3257	298.15	0.2691
323.15	0.2909	303.15	0.5670
333.15	0.2601	313.15	0.2309
343.15	0.2370	323.15	0.2123
353.15	0.2166	333.15	0.1924
363.15	0.2010	343.15	0.1752
373.15	0.1916	353.15	0.1600
393.15	0.1786	363.15	0.1484
Lemon juice		Pear juice	
303.15	0.2903	298.15	0.2687
313.15	0.2663	303.15	0.2725
323.15	0.2330	308.15	0.2696
333.15	0.2039	313.15	0.2643
343.15	0.1883	318.15	0.2538
353.15	0.1762	323.15	0.2435
363.15	0.1680	328.15	0.2308
373.15	0.1636	333.15	0.2179
383.15	0.1581	353.15	0.1828
393.15	0.1575	363.15	0.1840

As was shown by Breslau and Miller [155], this relation can be used to represent concentration dependence of the relative viscosity for concentrated aqueous electrolyte solutions if V_k is taken as an adjustable parameter. Abdulagatov et al. [122-127,140-143,152] used Eq. (40) to determine the values of V_k as a function of temperatures for some aqueous solutions from measured values of viscosity. Therefore, the relation (40) can be used to estimate the values of the hydrodynamic volume V_k (rigid molar volume) using the experimental relative viscosity data. Moulik and Rakshit [156] also used Eq. (40) to correlate the concentration dependence of the viscosity of 72 different aqueous solutions at high concentrations. Sahu and Behera [157] also represented experimental relative viscosity of aqueous solutions by extending the limiting Einstein equation as

$$\frac{\eta}{\eta_0} = 1 + 2.5\overline{V}x + k_1\overline{V}^2x^2 + k_2\overline{V}^3x^3 + k_3\overline{V}^4x^4, \tag{41}$$

where $\overline{V}$ is the volume of solution, x is the concentration. The measured viscosity values for pomegranate, pear, tangerine and lemon juices were fitted to the Eq. (40) (Magerramov et al. [14,15]). Note that this model is containing just one adjusting parameter V_k. The derived values of the parameter V_k for these juices are presented in Table 9 as a function of temperature.

4.3. Pressure Dependence of the Viscosity of Fruit Juices

Experimental viscosities of pomegranate and pear juices as a function of pressure are shown in Figure 10. The viscosity is little affected, (up to 1.5-1.7 %) at high temperatures (406 K) and up to 0.4-0.6 % at low temperatures (298 K), by pressure (at pressure changing between 0.1 and 10 MPa) along the constant temperature and constant concentration. The pressure dependence of the experimental viscosity of both juices in the range from 0.1 to 10 MPa is almost linear (see Figure 10). The values of experimental viscosity along an isotherm and constant concentration can be correlated within the experimental precision as a function of pressure by means of a linear expression (Kestin et al. [60], Kestin and Shankland [75])

$$\eta(T,x,P) = \eta_0(T,x)\left[1 + \beta(T,x)(P - P_0)\right], \tag{42}$$

where $\eta_0(T,x)$ is the viscosity at P_0 =0.101325 MPa (atmospheric pressure), can be correlated in terms of temperature and concentration, and

$\beta(T,x) = \frac{1}{\eta_0}\left(\frac{\partial \eta}{\partial P}\right)_{T,x}$ is the pressure coefficient of viscosity. This equation was successfully used by Kestin et al. [60] and Kestin and Shankland [75] to represent pressure dependence of viscosity of the aqueous electrolyte solutions.

4.4. Combined Effect of the Temperature and Concentration on the Viscosity of Fruit Juices

Both the temperature and concentration variations of the viscosity of fruit juices were combined by Rao [18] and Cepeda and Villarán [97] in a single exponential model (modification of the relation 31) for depectinised juice

$$\eta = \exp\left(a + \frac{b}{T} + cx\right) \text{ or } \eta = exp(a + cx)exp\left(\frac{b}{T}\right) \tag{43}$$

This relation was used by Jusczak and Fortuna [100] to represent measured values of viscosity of cherry juice. They also proposed other model (combination of the power and exponential functions) to combine the temperature and concentration dependence of viscosity

$$\eta = \eta_0 x^{\alpha} \exp\left(\frac{E_a}{RT}\right). \tag{44}$$

Rao et al. [112] and Ibarz et al. [107] used an exponential-type and a power-type relation to describe the effect of concentration on viscosity of fruit juices at constant temperature. Bayindirli [93,95] reported the model to describe the temperature and concentration effects on viscosity of grape juices as

$$\eta = \eta_0 \exp\left(\frac{(a + b/T)x}{100 - (c + dT)x}\right). \tag{45}$$

The experimental values of flow activation parameter E_a in Arrhenius equation for crab apple juice were fitted by Cepeda and Villarán [97] to a third degree polynomial equation. Alvarado and Romero [92] used exponential dependence on concentration of the parameter, $\eta_0 = \eta_1 \exp(ax)$ in Arrhenius equation in order to study of the combined effect of the temperature and concentration to the viscosity of juices. Ibraz et al. [108] proposed following form of the equation to describe the combined effect of temperature and concentration on the pear juice

$$\eta = K_2 \exp\left(\frac{E_a}{RT} + K_3 x + K_4 x^2\right)$$

or

$$\eta = K_2 \exp\left(K_3 x + K_4 x^2\right) \exp\left(\frac{E_a}{RT}\right) \tag{46}$$

Kaya and Sözer [114] and Altan and Maskan [184] proposed a simple equation for describing the combined effect of temperature and soluble solids content on the pomegranate juice

$$\eta = \eta_2 \exp\left(\frac{E_a}{RT} + Cx\right). \tag{47}$$

In order to describe the combined effect of temperature and concentration on the viscosity of fruit juices we used theoretically based Eqs. (32), (34) to (41) for the aqueous solutions. Because the theory cannot predict the concentration dependence of the parameters $\ln \eta_0$, b and c in Eqs. (31), (32) and temperature dependence of the parameters (V, E, A, B, C, D, V_k, and k_i) in Eqs. (34) to (41), we empirically extended these models to study of the combined effect of temperature and concentration on the viscosity of fruit juices by taken into account the concentration and temperature dependences of these parameters. These parameters (in Eqs. 32, 34 to 41) were considered as a simple polynomial function of concentration and temperature

$$\ln \eta_0 = \sum_{i=0}^{n} a_i x^i$$

and

$$b = \sum_{i=0}^{n} b_i x^i \, , \tag{48}$$

$$\mathrm{Y}(A,B,C,D,V_k,k_i) = \sum_{i=0}^{n} a_i T^i \, . \tag{49}$$

We experimentally studied the concentration and temperature dependences of the adjustable parameters in theoretically based Eqs. (31), (32), (34) to (41). As Tables 7 to 9 shows, the temperature dependences of theoretically and physically meaning parameters V_k, V, and E is very simple function of temperature and can be very readily described by the polynomial functions. The parameters V_k, V, and E are monotonically decreases with temperature and can be presented as simple polynomial function. In the first approaches these parameters can be presented as linear function of temperature (or first order polynomial function), although to more accurately representation of the experimental data in some cases high order polynomial functions (second, for example) are needed. Equations (31), (32), (34)-(41) together with (48) and (49) are providing the best description (within 0.8 to 2.0 %) of the experimental viscosity data for fruit juices.

The Arrhenius Eq. (31) with concentration dependences coefficients (Eq. 48) was applied to the measured values of the viscosity of pomegranate and pear juices. The derived values of the coefficients a_i and b_i in Eq. (48) are summarized in Table 10. Equation (31) together with (48) describes the experimental viscosity data for pomegranate and pear juices with accuracy of 2.0 %. Figure 14 compare the concentration dependence of the measured and calculated with various models the values of viscosity for tangerine and lemon juices at two selected temperatures of (303.15 and 363.15) K. As one can see from this figure, Eq. (34) and (37) excellent represents the present measurements for tangerine and lemon juices. These models can be recommended to future used for the describing of experimental viscosity data for liquid foods.

Table 10. Parameters a_i and b_i for Eq. (48).

Pomegranate juice (R2=0.998)					
	0	1	2	3	4
ai	1.0662×100	-1.0327×100	7.5160×10-2	-2.3096×10-3	2.4463×10-5
bi	6.7579×102	2.6653×102	-2.0716×101	0.6811×100	-7.4490×10-3
Pear juice (R2=0.998)					
	0	1	2	3	4
ai	-3.4380×100	-0.4434×100	2.3646×10-2	-4.175×10-4	0.0×100
bi	9.8756×102	1.6503×102	-8.894×100	0.1633×100	0.0×100

Table 11. Comparison accuracy and predictive capability of various combined models for the viscosity of lemon juice (Ki , a_i) parameters of the models).

	1	2	3	4	5	S*
$\eta = \eta_0 \exp\left(\frac{(a_1 + a_2/T)x}{100-(a_3+dT)x}\right)$						
ai	Ln η_0 =6.2208	225.596	-33317.26	30.54578	-0.015775	0.459×10-1
$\eta = K_2 \exp\left(\frac{E_a}{RT} + K_3 x + K_4 x^2\right)$						
Ki	LnK2=-6.9749	E_a/R =2167.48	K3=0.008528	K4=0.0005404	-	0.534×10-2
$\eta = \eta_0 \exp\left(\frac{b}{T}\right),\ b = \sum_{i=0}^{4} a_i x^i$						
ai	-6.2173	0.002522	1745.58	13.6980	-	0.583×10-2
$\eta = \exp\left(a_1 + \frac{a_2}{T} + a_3 x\right),\ \eta = \eta_2 \exp\left(\frac{E_a}{RT} + Cx\right)$						
	1	2	3	4	5	S*
ai	Ln a_1 =-7.4375	2167.48	0.04214	-	-	0.699×10 -2
$\eta = \eta_0 x^{\alpha} \exp\left(\frac{E_a}{RT}\right)$						
ai	Ln η_0 =-10.171	E_a/R =2167.48	1.1982	-	-	0.152×10 -1
$\eta = \eta_0 \exp[f(x)],\ f(x) = a_1 x + a_2 x^2$						
ai	0.02352	0.0003062	-	-	-	0.216×10 -1

Table 11. (Continued).

	1	2	3	4	5	S*
$\frac{\eta}{\eta_0} = 1 + a_1 x + a_2 x^2 + a_3 x^3$						
ai	0.08634	-0.004037	0.0001075	-	-	0.515×10-2
$\frac{\eta}{\eta_0} = 1 + 2.5\phi + 10.05\phi^2 = 1 + 2.5V_k x + 10.05V_k^2 x^2$						
	1	2	3	4	5	S*
V_k	0.01043	-	-	-	-	0.759×10-2
$\frac{\eta}{\eta_0} = 1 + 2.5\bar{V}x + a_1\bar{V}^2 x^2 + a_2\bar{V}^3 x^3 + a_3\bar{V}^4 x^4$						
ai	0.03693	-0.00625	0.0002126	-0.00000165	-	0.524×10-2
$\frac{\eta}{\eta_0} = 1 + a_1\sqrt{x} + a_2 x + a_3 x^2$						
ai	0.717976	-0.206593	0.004243	-	-	0.510×10-2
$\eta = \frac{\eta_0 \exp(xE)}{1 + xV}$						
E,V	0.09178	-0.10738	-	-	-	0.130×10-1
$\eta = \eta_0 \exp\left(\frac{b}{T} + \frac{c}{T^2}\right)$, $ln\eta_0 = a_1 x$, $b = a_2 + a_3 x$, $c = a_4 + a_5 x$						
ai	0.09358	-2528.299	-49.8041	729457.81	10993.444	0.319×10-2

* Error function $S = \frac{1}{N}\sum_i \left(Y_i^{cal} - Y_i^{\exp}\right)$

Table 11 is providing the comparison accuracy of the models for lemon juice. As one can see from the table, Eqs. (37) and (39) are provide the best representation of the measured values of the viscosity of lemon juice. These models based on the pure water viscosity and contain only 3 adjustable parameters. Reasonable representation of the temperature and concentration effect on the viscosity of lemon juice was observed also for the one-parametric model (40) which based on the extended Einstein relation. The predictive (extrapolating properties) capability of the model (Eq. 40) is much better than other models.

Bálint [208] developed combined model by using the combination polynomial and power functions for the viscosity of whey

$$\eta = 0.0638(9/5t + 32)^{0.9982}\left(1.0042 + 3.11x + 14.7x^2\right). \qquad (50)$$

Recently, an artificial network (ANN) model was developed for the prediction of viscosity of fruit juices as function of concentration and temperature by Rai et al. [220]. The accuracy of the model is within 3.78 to 4.20 mPa·s. Peacock [211] proposed the another model for the viscosity of sucrose solutions in the form

$$\log_{10}\eta = a_0 N - a_1(30 - T)\left(a_2 + a_3 N^{1.25}\right)/(91 + T), \qquad (51)$$

where $N = x(1900 - 18x)$. The range applicability of the model is from 10 to 80 °C and between 0 and 85 °Brix.

Chapter 5

CONCLUSION

In this chapter the different techniques employed for the measurement of viscosity of liquids and liquid foods were discussed. These were the capillary-flow technique, the oscillating-disc technique and the falling-body method. From the discussion of the techniques it is apparent that measurements performed with the capillary-flow viscometer and the oscillating-disc viscometer enjoy a low degree of uncertainty. Hence, measurements would be expected to attain an uncertainty of better than 2 %. Measurements performed with the falling-body viscometer would be expected to attain a higher uncertainty.

The temperature, pressure, and concentration dependences of the viscosity of fruit juices were studied. The values of the flow activation energy E_a of the temperature dependence Arrenius equation were calculated for the viscosity of pomegranate, pear, tangerine, and lemon juices as a function of concentration and pressure. In order to represent combined effect of concentration and temperature on the viscosity of fruit juices the various models were examined.

A new correlation and prediction models for the viscosity was developed. The applicability of the various theoretical based models used previously for aqueous solutions to describe the effect of temperature and concentration on viscosity of fruit juices was studied. The measured values of viscosity of fruit juices were used to test and confirm the accuracy and predictive capability of various theoretical and semi-empirical viscosity models.

An accuracy and predictive capability of the semitheoretical models for the viscosity of fruit juices proposed in this work are much better than various multiparametric correlations. It was found that the prediction models (Eq. 31 to 41) developed in this work for the viscosity of fruit juices can be adopted with satisfaction. The models (Eq. 32,34,35,37, and 40) with temperature and concentration depending parameters are accurately represent the combined effect of concentration and temperature on the viscosity of fruit juices in wide temperature and concentration ranges and can be recommended for the future scientific and engineering used.

APPENDIX
EXPERIMENTAL VISCOSITY DATA FOR FRUIT JUICES

Table 1A. Experimental viscosity (mPa·s) of natural pomegranate juice as a function of temperature and pressure for 11.0 °Brix.

T (K)	0.1 MPa	5 MPa	10 MPa
292.95	1.642	-	1.646
296.85	1.480	-	-
301.05	1.318	-	-
304.55	1.213	-	1.216
308.05	1.119	-	-
311.55	1.032	1.034	-
316.55	0.928	-	0.932
321.85	0.837	-	-
325.95	0.777		0.782
328.95	0.740	-	-
332.35	0.699	-	0.704
337.95	0.637	-	-
343.05	0.587	-	0.593
348.15	0.541	-	-
353.05	0.500	0.502	-
358.15	0.465	-	-
362.95	0.435	-	0.440
368.45	0.405	-	-
374.85	0.374 [a]	-	0.379
379.95	0.352 [a]	0.354	-
384.85	0.332 [a]	-	0.337
388.15	0.321 [a]	-	-
394.75	0.299 [a]	-	0.304
402.95	0.275 [a]	0.277	0.279

[a]At pressure of 0.3 MPa.

Table 2A. Experimental viscosity (mPa·s) of pomegranate juice as a function of temperaturte and concentration.

T (K)	23 °Brix	30 °Brix	35 °Brix	40 °Brix
293.15	2.017	3.307	4.968	7.082
298.15	1.765	2.828	4.248	5.994
303.15	1.555	2.488	3.593	5.081
313.15	1.231	1.902	2.659	3.702
323.15	1.001	1.498	2.051	2.863
333.15	0.827	1.200	1.606	2.252
343.15	0.697	0.985	1.288	1.778
353.15	0.597	0.817	1.055	1.443
363.15	0.533	0.692	0.875	1.192

Table 3A. Experimental viscosity (mPa·s) of pear juice as a function of temperature and pressure for 15.2 °Brix.

T (K)	0.1 MPa	5 MPa	10 MPa
294.10	1.975	1.977	1.979
298.60	1.781	-	1.784
305.12	1.550	-	1.554
311.60	1.342	-	1.346
318.44	1.154	1.157	-
325.04	1.009	1.011	-
332.40	0.872	0.875	-
340.00	0.757	-	0.762
349.71	0.647	-	0.651
358.63	0.574	-	0.579
367.23	0.523	0.526	0.528
374.80	0.488[a]	-	0.493
382.50	0.460[a]	-	0.465
391.30	0.427[a]	-	0.432
402.71	0.381[a]	-	0.386

[a] At pressure of 0.3 MPa.

Table 4A. Experimental viscosity (mPa·s) of pear juice as a function of temperature and concentration.

T (K)	20 °Brix	25 °Brix	30 °Brix
298.15	1.902	2.194	3.061
303.15	1.721	1.990	2.749
308.15	1.551	1.795	2.438
313.15	1.403	1.621	2.162
318.15	1.252	1.448	1.921
323.15	1.122	1.303	1.703
328.15	1.012	1.164	1.510
333.15	0.920	1.040	1.341
343.15	0.761	0.861	1.091
353.15	0.646	0.725	0.902
363.15	0.590	0.655	0.782

Table 5A. Experimental viscosity (mPa·s) of tangerine juice as a function of temperature and concentration.

T (K)	15 °Brix	21 °Brix	26°Brix	35°Brix	40°Brix
303.15	1.608	1.904	2.420	4.608	6.627
313.15	1.262	1.482	1.887	3.356	4.695
323.15	1.010	1.179	1.502	2.540	3.419
333.15	0.818	0.954	1.219	1.966	2.600
343.15	0.677	0.791	1.006	1.580	2.074
353.15	0.567	0.658	0.840	1.304	1.696
363.15	0.477	0.556	0.712	1.107	1.435
373.15	0.412 a	0.482	0.620	0.966	1.262
383.15	0.357 a	0.423	0.548	0.862	1.127
393.15	0.315 a	0.375	0.492	0.771	1.020

[a] At pressure of 0.3 MPa.

Table 6A. Experimental viscosity (mPa·s) of lemon juice as a function of temperature and concentration.

T, K	17°Brix	22°Brix	30°Brix	40°Brix	45°Brix
303.15	1.563	1.816	2.571	4.599	6.600
313.15	1.233	1.423	2.106	3.390	4.710
323.15	0.999	1.133	1.596	2.547	3.433
333.15	0.832	0.922	1.286	1.983	2.430
343.15	0.699	0.763	1.056	1.587	2.030
353.15	0.595	0.635	0.885	1.325	1.729
363.15	0.510	0.543	0.764	1.146	1.480
373.15	0.444 a	0.475	0.677	1.015	1.310
383.15	0.393 a	0.420	0.607	0.910	1.176
393.15	0.350 a	0.378	0.549	0.822	1.060

[a] At pressure of 0.3 MPa.

REFERENCES

[1] Moressi, M.; Spinosi, M. Engineering factors in the production of concentrated fruit juices.1. Fluid physical properties of orange juices. *J. Food. Technol.* 1980, *15*, 265-276.

[2] Crandall, P.G.; Chen, C.S.; Carter, R.D. Models for predicting viscosity of orange juice concentrate. *Food Technology* 1982, *36*, 245-252.

[3] Famelart, M.-H.; Chapron, L.; Piot, M.; Brule, G.; Durier, C. High pressure –induced gel formation of milk and whey concentrates. *J. Food Eng.* 1998, *36*, 149-164.

[4] Farkas, D.F.; Hoover, D.G. High pressure processing. *J. Food Sci. Supplem.* 2000, *65*, 47-64.

[5] Farr, D. High pressure technology in the food industry. Trends in Food Science and Technology 1990, 14-16.

[6] Felipe, X.; Capellas, M.; Law, A.J.R. Comparison of the effects of high-pressure treatments and heat pasteurization on the whey proteins in goat's milk. *J. Agr. Food Chem.* 1997, *45*, 627-631.

[7] Delza, R.; Rosenthal, A.; Abadio, F.B.D.; Silva, C.H.O.; Castillo, C. Application of high pressure technology in the fruit juice processing: benefits perceived by consumers. *J. Food Eng.* 2005, *67*, 241-246.

[8] Desrumaux, A.; Marcand, J. Formation of sunflower oil emulsions stabilized by whey proteins with high pressure homogenization (up to 350 MPa): effect of pressure on emulsion characteristics. *Int. J. Food Sci. and Techn.* 2002, *37*, 263-269.

[9] Garcia, F.; Butz, P.; Bognar, A.; Tauscher, B. Antioxidative capacity, nutrient content and sensory quality of orange juice and an orange-lemon-carrot juice product after high pressure treatment and storage in different packaging. *European Food Res. Tech.* 2001, *213*, 290-296.

[10] Magerramov, M.A. Density of the concentrates of peach and pomegranate juices at elevated state parameters. *J. Eng. Phys. Thermophys.* 2007, *79*, 811-816.

[11] Magerramov, M.A.; Abdulagatov, A.I.; Azizov, N.D.; Abdulagatov, I.M. *J. Food and Bioprocess Technology: An International Journal* 2007, in press.

[12] Magerramov, M.A.; Abdulagatov, A.I.; Azizov, N.D.; Abdulagatov, I.M. Thermal conductivity of peach, raspberry, cherry, and plum juices as a function of temperature and concentration. *J. Food Proc. Eng.* 2006, *29*, 304-326.

[13] Magerramov, M.A.; Abdulagatov, A.I.; Azizov, N.D.; Abdulagatov, I.M. Thermal conductivity of pear, sweet-cherry, apricot, and cherry-plum juices as a function of temperature and concentration. *J. Food Sci. E.-Food Eng. and Phys. Properties* 2006, *71* 238-244.

[14] Magerramov, M.A.; Abdulagatov, A.I.; Azizov, N.D.; Abdulagatov, I.M. Effect of temperature, concentration, and pressure on the viscosity of pomegranate and pear juice concentrates. *J. Food Engineering* 2007, *80*, 476-489.

[15] Magerramov, M.A.; Abdulagatov, A.I.; Azizov, N.D.; Abdulagatov, I.M. Viscosity of tangerine and lemon juices as a function of temperature and concentration. *Int. J. Food Sci. and Technology* 2007, *42*, 804-819.

[16] Quashou, M.S.; Vachon, R.I.; Touloukian, Y.S. *Thermal conductivity of foods*. Part 1; ASHRAE Transactions: New York, NY, 1972; Vol. 78, pp 165-182.

[17] 17. Rha, C. *Theory, Determination and Control of Physical Properties of Food Materials*; D. Riedel Pub. Co.: Dordrecht, Holland, 1975.

[18] Rao, M.A. Viscoelastic properties of fluid and semi solid foods. In: *Physical and Chemical Properties of Food*; Editor M.R. Okos; ASAE: New York, NY, 1986; 14-34.

[19] Rao, M.A. *Rheology of Fluid and Semisolid Foods. Principles and Applications*; Aspen Publishers, Inc.: Gaithersburg, Maryland, 1999.

[20] Rahman, *Food Properties Hand Book*; CRC Press: Boca Raton, FL, 1995.

[21] Cuevas, R.; Cheryan, M. Thermal conductivity of liquid foods-review. *J. Food Proc. Eng.* 1979, *2*, 283-306.

[22] Chen, C.S. Physical and rheological properties of juices. In: *Fruit Juice Processing Technology*; Editor S. Nagy; C.S. Chen; Ph.E. Shaw; Agscience, Inc.: Auburndale Florida, Chapter 3, 1993; pp 56-82.

[23] Choi, I.; Okos, M.R. The thermal properties of liquid foods. Paper no 83-6516. *Winter Meeting*; ASAE: Chicago, IL, 1983.

[24] Choi, Y.; Okos, M. Effect of temperature and composition on the thermal properties of foods. In: *Food Engineering and Process Applications, Transport Phenomena*; Editors M.L. Maguer, P. Jelen; EASP: New York, NY, 1986; Vol.1, pp 93-101.

[25] Choi, Y.; Okos, M. Thermal properties of liquid foods-Review. In: *Physical and Chemical Properties of Food*; Editor M.R. Okos; ASAE: New York, NY, 1986; pp 35-77.

[26] Bourne, M.C. *Food Texture and Viscosity: Concept and Measurement*; Academy Press: New York, NY, 1982.

[27] Sweat, V.E. Thermal properties of foods. In: *Engineering Properties of Foods*; Editors M.A. Rao; S.S.H. Rizvi; Marcel Dekker, Inc.: New York, NY, Chapter 1,1986; pp 49-87.

[28] Steffe, J.F.; Mohamed, I.O.; Ford, E.W. Rheological properties of fluid foods: Data compilation. In: *Physical and Chemical Properties of Food*; Editor M.R. Okos; ASAE: New York, NY, 1986; pp 1-13.

[29] Saravacos, G.D.; Maroulis, Z.B. *Transport Properties of Foods*; Marcel Dekker, Inc.: New York, NY, 2002.

[30] Polley, S.L.; Snyder, O.P.; Kotur, P. A compilation of thermal properties of foods. *Food Technology* 1980, *34*, 76-108.

[31] Padmanabhan, M. Measurement of extensional viscosity of viscoelastic liquid foods. *J. Food Eng.* 1995, *25*, 311-327.

[32] Krokida, M.K.; Maroulis, Z.B.; Saravacos, G.D. Rheological properties of fluid fruit and vegetable puree products: Compilation of literature data. *Int. J. Food Properties* 2001, *4*, 179-200.

[33] Golubev, I.F. *Viscosity of Gases and Gaseous Mixtures*; Phys. Math. Press: Moscow, Russia, 1959.

[34] Golubev, I.F.; Agaev, N.A. *Viscosity of Hydrocarbons*; Azernashr: Baku, Azerbaijan, 1964.

[35] Golubev, I.F.; Gnesdilov, N.E. *Viscosity of Gaseous Mixtures*; GSSSD: Moscow, Russia, 1971.

[36] Guseinov, K.D. Study of the thermodynamic and transport properties of some organic liquids in the wide range of parameters of state. *Ph. D. Thesis*, API: Baku, Azerbaijan, 1979.

[37] Abdulagatov, I.M.; Rasulov, S.M. Viscosity of n-pentane and n-heptane and their mixtures within the temperature range from 298 K up to critical points at the saturation vapor pressure. *Ber. Bunsenges. Phys. Chem.* 1996, *100*, 148-154.

[38] Saravacos, G.D. Effect of temperature on viscosity of fruit juices and purees. *J. Food Science* 1970, *35*, 122-125.

[39] Rouse, A.H.; Albrigo, L.G.; Huggart, R.L.; Moore, E.L. Viscometric measurements and pectic content of frozen concentrated orange juices for citrus futures. *Ann. Meeting of the Florida State Horticultyral Society* 1974, *87*, 293-296.

[40] Rao, M.A.; Anantheswaran, R.C. Rheology of fluids in food processing's. *Food Tech.* 1982, *36*, 116-126.

[41] Wakeham, W.A.; Nagashima, A.; Sengers, J.V. (Editors) *Measurements of the Transport Properties of Fluids*; Blackwell Scientific Publications: Oxford, Vo. III, 1991.

[42] Hagenbach, S. *Ann. Phys., Lpz.* 1860, *109*, 385-426.

[43] Barr, G. *A Monograph of Viscosity*; Oxford University Press: London, 1931.

[44] Kestin, J.; Sokolov, M.; Wakeham, W.A. Theory of capillary viscometers. *Appl. Sci. Res.* 1973, *27*, 241-264.

[45] Van Wazer, J.R.; Lyons, J.W.; Kim, K.Y.; Colwell, R.E. *Viscosity and Flow Measurements*; Intersciences: New York, NY, 1963.

[46] Abdulagatov, I.M.; Azizov, N.D. Viscosity for Aqueous Li_2SO_4 Solutions at Temperatures from 298 to 575 K and at Pressures up to 30 MPa. *J. Chem. Eng. Data* 2003, *48*, 1549-1556.

[47] Rivkin, S.L.; Kremenevskaya, E.A.; Romashin, S.N.; Tractuev, O.M. Experimental investigation of density and viscosity of aqueous boron solutions. In: *Proc. 10th Int. Conf. Prop. Steam*; Editors, V.V. Sytchev; A.A. Aleksandrov; Mir Publ.: Moscow, 1984; Vol.2, pp 289-298.

[48] Puchkov, L.V.; Sargaev, P.M. Viscosity of lithium, sodium, potassium and ammonium nitrates solutions at temperature up to 275 °C. *Russ. J. Appl. Chem.* 1973, *46*, 2637-2640.

[49] Semenyuk, E.N.; Zarembo, V.I.; Feodorov, M.K. Experimental apparatus for the measurements of viscosity of aqueous solutions at temperatures 273-673 K and at pressures up to 200 MPa. *Russ. J. Appl. Chem.* 1977, *50*, 315-319.

[50] Mashovets, V.P.; Puchkov, L.V.; Sargaev, P.M.; Feodorov, M.K. Experimental apparatus for the viscosity measurements of aqueous solutions at temperatures up to 275 °C. *Russ. J. Appl. Chem.* 1971, *44*, 90-94.

[51] Pepinov, R.I.; Yusufova, V.J.; Lobkova, N.V.; Panakhov, I.A. Viscosity of aqueous potassium chloride solutions over a wide range of state

parameters. In: *Proc. 10th Int. Conf. Prop. Steam*; Editors; V.V. Sytchev; A.A. Aleksandrov, Mir Publ.: Moscow, 1984; Vol. 2, pp 196-202.

[52] Kestin, J.; Sengers, J.V.; Kamgar-Parsi, B.; Levelt Sengers, J.M.H. Thermophysical properties of fluid H_2O. *J. Phys. Chem. Ref. Data* 1984, *13*, 175-189.

[53] *International Organization of Standardization (IOS) 4787* Laboratory glassware-Volumetric glassware-Methods for use and testing capacity, IOS: Geneve, 1994.

[54] Coleman, HW; Steele, WG *Experimentation and Uncertainty Analysis for Engineers*; John Wiley and Sons: New York, NY, 1989.

[55] DiPippo, R.; Kestin, J.; Whitelaw, J.H. A high-temperature oscillating-disk viscometer. *Physica* 1966, *32*, 2064-2080.

[56] Kestin, J.; Paul, R.; Shankland, I.R.; Khalifa, H.E. A high-temperature, high-pressure oscillating –disk viscometer for concentrated ionic solutions. *Ber. Bunsenges. Phys. Chem.* 1980, *84*, 1255-1260.

[57] Kestin, J.; Khalifa, H.E.; Ro, S.T.; Wakeham, W.A. Preliminary data on the pressure effect on the viscosity of sodium chloride-water solutions in the range 10-40 °C. *J. Chem. Eng. Data* 1977, *22*, 207-214.

[58] Kestin, J.; Khalifa, H.E.; Sookiazian, H.; Wakeham, W.A. Experimental investigation of the effect of pressure on the viscosity of water in the temperature range 10-150 °C. *Ber. Bunsenges. Phys. Chem.* 1978, *82*, 180-188.

[59] Kestin, J.; Khalifa, H.E.; Abe, Y.; Grimes, C.E.; Sookiazian, H.; Wakeham, W.A. Effect of pressure on the viscosity of aqueous sodium chloride solutions in the temperature range 20-150 °C. *J. Chem. Eng. Data* 1978, *23*, 328-336.

[60] Kestin, J.; Shankland, I.R.; Paul, R. The viscosity of aqueous KCl solutions in the temperature range 25–200°C and the pressure range 0.1–30 MPa. *Int. J. Thermophys.* 1981, *2*, 301-314.

[61] Kestin, J.; Shankland, I.R. A re-evaluation of the relative measurements of viscosity with oscillating –disk viscometers. *J. Appl. Math. Phys.* 1981, *32*, 533-545.

[62] Correia, R.J.; Kestin, J.; Khalifa, H.E. Measurements and calculation of the viscosity of mixed aqueous solutions of NaCl and KCl in the temperature range 25-150 C and the pressure range 0-30 MPa. *Ber. Bunsenges. Phys. Chem.* 1979, *83*, 20-24.

[63] Kestin, J.; Khalifa, H.E. Measurement of logarithmic decrements through the measurement of time. *Appl. Sci. Res.* 1976, *32*, 483-493.

[64] Newell, G.F. Theory of oscillation type viscometers V: Disk oscillating between fixed plates. *Z. Angew. Math. Phys.* 1959, *10*, 160-174.

[65] Kestin, J.; Wang, H.E.; Providence, R.I. Corrections for the oscillating-disk viscometer. *J. Appl. Mech.* 1957, *79*, 197-206.

[66] Kestin, J.; Newell, G.F. Theory of oscillation type viscometers: The oscillating cup Part I. *Z. Angew. Math. Phys.* 1957, *8*, 433-442.

[67] Kestin, J.; Moszynski, J.R.; Providence, R.I. An Instrument for the measurement of the viscosity of steam and compressed water. *Trans. ASME* 1958, *80*, 1009-1014.

[68] Kestin, J.; Leidenfrost, W.; Liu, C.Y. Z. On relative measurements of the viscosity of gases by the oscillating-disk method. *Angew. Math. Phys.* 1959, *10*, 558-564.

[69] Nieuwoudt, J.C.; Kestin, J.; Sengers, J.V. On the theory of oscillating-body viscometers. *Physica A* 1987, *142*, 53-74.

[70] Nieuwoudt, J.C. An extension of the theory of oscillating cup viscometers. *Int. J. Thermophys.* 1990, *11*, 525-535.

[71] Tørklep, K.; Øye, H.A. Viscostiy of molten alkaline-earth chlorides. *J. Chem. Eng. Data* 1982, *27*, 387-391.

[72] Iwasaki, H.; Kestin, J. The viscosity of argon-helium mixtures at 20 °C and 1, 25, and 50 atm.. *Physica* 1963, *29*, 1345-1372.

[73] Kestin, J.; Sokolov, M.; Wakeham, W.A.Viscosity of liquid water in the range -8 °C to 150 °C. *J. Phys. Chem. Ref. Data* 1978, *7*, 941-948.

[74] Grimes, C.G.; Kestin, J.; Khalifa, H.E. Viscosity of aqueous potassium chloride solutions in the temperature range 25-150°C and the pressure range 0-30 MPa. *J. Chem. Eng. Data* 1979, *24*, 121-126.

[75] Kestin, J.; Shankland, I.R. Viscosity of aqueous NaCl solutions in the temperature range 25–200 °C and in the pressure range 0.1–30 MPa. *Int. J. Thermophys.* 1984, *5*, 241-263.

[76] Kawata, M.; Kurase, K.; Nagashima, A.; Yoshida, K. In: *Experimental Thermodynamics. III. The Measurement of Transport Properties of Fluids*; Editors, W.A. Wakeham; A. Nagashima; J.V. Sengers; Blackwell Scientific: Oxford, 1991; pp 49-75.

[77] Faxen, H. Die bewegung einer starren Kugel längs der Achse eines mit zäher Flüssigkeit gefüllten Rohres. *Arkiv f. Mat. Astron. Och Fysik* 1923, *17*, 1-28.

[78] Fidleris, V.; Whitemore, R.L. experimental determination of the wall effect for spheres falling axially in cylindrical vessels. *Brit. J. Appl. Phys.* 1961, *12*, 490-494.

[79] Sutterby, J.L. Falling sphere viscometer. *Trans. Soc. Rheol.* 1973, *17*, 559-573.

[80] Oseen, C.W. Uber. Die electromagnetischen schwingungen an dunnen ringen. *Arkiv f. Mat. Astron. Och Fysik* 1913, *9*, 1-15.

[81] Huner, B.; Hussey, R.G. Cylinder drag at low Reynolds number. *Phys. Fluids* 1977, *20*, 1211-1218.

[82] Maxworthy, T. Accurate measurements of sphere drag at low Reynolds numbers. *J. Fluid Mech.* 1965, *23*, 369-372.

[83] Proudman, I.; Pearson, J.R.A. Expansions at small Reynolds numbers for the flow past a sphere and a circular cylinder. *J. Fluid Mech.* 1957, *2*, 237-262.

[84] Maude, A.D. End effects in a falling-sphere viscometer. *Brit. J. Appl. Phys.* 1961, *12*, 293-295.

[85] Flude, M.J.C.; Daborn, J.E. Viscosity measurement by means of falling spheres compared with capillary viscometry. *J. Phys. E.* 1982, *15*, 1313-1321.

[86] Tanner, R.I. End effects in falling-ball viscometry. *J. Fluid Mech.* 1963, *17*, 161-170.

[87] Isdale, J.D.; Easteal, A.J.; Wolf, L.A. Shear viscosity of methanol and methanol + water mixtures under pressure. *Int. J. Thermpohys.* 1985, *6*, 439-450.

[88] Tanaka, Y.; Matsuda, Y.; Fujuwara, H. Viscosity of (water + alcohol) mixtures under high pressure. *Int. J. Thermpohys.* 1987, *8*, 147-163.

[89] Melikov, R.A. *Ph.D. Thesis*, Azerbaijan Power Engineering Research Institute, Baku, 1996.

[90] Melikov, R.A. In: *Scientific Transactions of Azerbaijan State Oil Academy. Power Engineering of Oil and Gas Industry* 1999, *4*, 144-147.

[91] Melikov, R.A. In: *Proc. Conference, "Ratsional'noe Ispol'zovanie Energo-Resursov I Nadezhnost Electrooborudovanii"*, Azerbaijan State Oil Academy: Baku, Azerbaijan, 2000; pp 360-363.

[92] Alvarado, J.D.; Romero, C.H. Physical properties of fruits I-II. Density and viscosity of juices as functions of soluble solids content and temperature. *Latin Amer. Applied Research* 1989, *19*, 15-21.

[93] Bayindirli, L. Mathematical analysis of variation of density and viscosity of apple juice with temperature and concentration. *J. Food Processing and Preservation* 1992, *16*, 23-28.

[94] Bayindirli, L.; Özsan, O. Modeling of thermophysical properties of sourchery juice. *Gida* 1992, *17*, 405-407.

[95] Bayindirli, L. Density and viscosity of grape juice as a function of concentration and temperature. *J. Food Processing and Preservation* 1993, *17*, 147-151.

[96] Bayindirli, L.; Sahin, S.; Artik, N. The effect of clarification methods on pomegranate juice quality. *Fruit Processing* 1994, *4*, 267-270.

[97] Cepeda, E.; Villarán, M.C. Density and viscosity of Malus floribunda juice as a function of concentration and temperature. *J. Food Eng.* 1999, *41*, 103-107.

[98] Constenla, D.T., Lozano J.E.; Crapiste, G.H. Thermophysical properties clarified apple juice as a function of concentration and temperature. *J. Food Science* 1989, *54*, 663-668.

[99] Juszczak, L.; Fortuna, T. Viscosity of concentrated strawberry juice. Effect of temperature and soluble solids content. *J. Polish Agr. University* 2003, *6*, 1-7.

[100] Juszczak, L.; Fortuna, T. Effect of temperature and soluble solids content on the viscosity of cherry juice concentrate. *Int. Agrophys.* 2004, *18*, 17-21.

[101] Ibarz, A.; Gonzalez, C.; Esplugas, S. Rheology of clarified fruit juices. III: Orange juices. *J. Food Eng.* 1994, *21*, 485-494.

[102] Ibarz, A.; Garvin, A.; Costa, J. Rheological behavior of sloe (*prunus spinosa*) fruit juices. *J. Food Eng.* 1996, *27*, 423-430.

[103] Ibarz, A.; Garvin, A.; Costa, J. Rheological behavior of loquat (*eriobotrya japonica*) juices. *J. Food Eng.* 1996, *27*, 175-184.

[104] Ibarz, A.; Pagán, J. Rheology of raspberry juices. I: *J. Food Eng.* 1987, *6*, 269-289.

[105] Ibarz, A.; Gonzalez, C.; Esplugas, S.; Vicente, M. Rheology of clarified fruit juices. I: Peach juices. *J. Food Eng.* 1992, *15*, 49-61.

[106] Ibarz, A.; Pagán, J.; Miguelsanz, R. Rheology of clarified fruit juices. II: Blackcurrant juices. *J. Food Eng.* 1992, *15*, 63-73.

[107] Ibarz, A.; Vicente, M.; Graell, J. Rheological behavior of apple juice and pear juice and their concentrates. *J. Food Eng.* 1987, *6*, 257-267.

[108] Ibarz, A.; Pagan, J.; Gutierrez, J.; Vicente, M. Rheological properties of clarified pear juices concentrates. *J. Food Eng.* 1989, *10*, 57-63.

[109] Barros, S.M. *Quality summary. Citrus juice products from FMC citrus juice extractors*; Citrus Machinery Div., FMC Corp.: Lakeland, Fla., 1977; 1978; 1979.

[110] Rao, M.A.; Cooley, H.J.; Ortloff, C.; Chung, K.; Wijts, S.C. Influence of rheological properties of fluid and semisolid foods on the performance of a filler. *J. Food Proc. Eng.* 1993, *16*, 289-304.

[111] Rao, K.L.; Eipeson, W.E.; Rao, P.N.S.; Patwardhan, M.V.; Ramanathan, P.K. Rheological properties of mango pulp and concentrate. *J. Food Sci. Tech.* 1985, *22*, 30-33.

[112] Rao, M.A.; Cooley, H.J.; Vitali, A.A. Flow properties of concentrated juices at low temperatures. *Food Technology* 1984, *38*, 113-119.

[113] Vitali, A.A.; Rao, M.A. Flow properties of low-pulp concentrated orange juice: Effect of temperature and concentration. *J. Food Sciences* 1984, *49*, 882-888.

[114] Kaya, A.; Sözer, N. Rheological behavior of sour pomegranate juice concentrates (*Punica granatum L.*). *Int. J. Food Science and Technology* 2005, *40*, 223-227.

[115] Kaya, A. Rheology of mulberry pekmez. *J. Texture Studies* 2001, *32*, 335-342

[116] Saravacos, G.D.; Kostaropoulos, A.E. Transport properties in processing of fruits and vegetables. *Food Technology*, September, 1995, 99-105.

[117] Saravacos, G.D. In: *Engineering Properties of Foods*. Editors; M.A. Rao; S.S.H. Rizvi; Marcel Dekker, Inc.: 1986, pp 89-132.

[118] Glasstone, S.; Laidler, K.; Eyring, E. *Theory of Rate Processes*; McGraw-Hill: New York, NY, 1941.

[119] Erday-Grúz, T. *Transport Phenomena in Aqueous Solutions*. John Wiley and Sons Inc.: New York, NY, 1974.

[120] Millat, J.; Dymond, J.H.; Nieto de Castro, C.A. Editors, *Transport Properties of Fluids. Their Correlation, Prediction and Estimation*; IUPAC, Cambridge University Press: New York, NY, 1996.

[121] Stokes, R.H.; Mills, R. *Viscosity of Electrolytes and Related Properties*; Pergamon Press, Inc.: New York, NY, 1965.

[122] Abdulagatov, I.M.; Azizov, N.D. Viscosity of aqueous LiI solutions at 293 - 525 K and 0.1-40 MPa. *Thermochimica Acta* 2005, *439*, 8-20.

[123] Abdulagatov, I.M.; Azizov, N.D. Viscosity of aqueous calcium chloride solutions at high temperatures and high pressures. *Fluid Phase Equilibria* 2006, *240*, 204-219.

[124] Abdulagatov, I.M.; Azizov, N.D. Experimental study of the effect of temperature, pressure, and concentration on the viscosity of aqueous NaBr solutions. *J. Sol. Chem.* 2006, *35*, 705-738.

[125] Abdulagatov, I.M.; Azizov, N.D. Experimental study of the effect of temperature, pressure, and concentration on the viscosity of aqueous $SrCl_2$ solutions. *J. Chem. Eng. Data* 2007, *52*, 841-850.

[126] Abdulagatov, I.M.; Azizov, N.D. Experimental study of the effect of temperature, pressure, and concentration on the viscosity of aqueous KBr. *J. Sol. Chemistry* 2007, in press.

[127] Abdulagatov, I.M.; Zeinalova, A.B.; Azizov, N.D. Viscosity of aqueous $Ni(NO_3)_2$ solutions at temperatures from (297 to 475) K and at pressures up to 30 MPa and concentration between (0.050 and 2.246) $mol \cdot kg^{-1}$. *J. Chem. Thermodyn.* 2006, *38*, 179-189.

[128] Chirife, J. Microbial growth at reduced water activities: some physico-chemical properties of compatible soluties. *J. Appl. Bacteriol.* 1984, *56*, 259-268.

[129] Leyenndekkers, J.V.; Hunter, R.J. Refractive index of aqueous electrolyte solutions. Extrapolation to other temperatures, pressures, and wavelengths and to multicomponent systems. *J. Chem. Eng. Data* 1977, *22*, 427-431.

[130] Leyenndekkers, J.V.; Hunter, R.J. The viscosity of aqueous electrolyte solutions and the TTG model. *J. Phys. Chem.* 1977, *81*, 1657-1663.

[131] Leyenndekkers, J.V. The viscosity of aqueous electrolyte solutions and the TTG model. *J. Sol. Chem.* 1979, *8*, 853-869.

[132] Goldsack, D.E.; Franchetto, R.C. The viscosity of concentrated electrolyte solutions. I. Concentrated dependence at fixed temperature. *Can. J. Chem.-Rev. Can. Chim.* 1977, *55*, 1062-1072.

[133] Goldsack, D.E.; Franchetto, R.C. The viscosity of concentrated electrolyte solutions. II. Temperature dependence. *Can. J. Chem.-Rev. Can. Chim.* 1978, *56*, 1442-1450.

[134] Chirife, J.; Buera, M.P. A simple model for predicting the viscosity of sugar and oligosaccharide solutions. *J. Food Eng.* 1997, *33*, 211-226.

[135] Falkenhagen, H.; Dole, M. Die innere reibung von elektrolytischen losungen und ihre deutung nach der Debyeschen theorie. *Phys. Z.* 1929, *30*, 611-622.

[136] Onsager, L.; Fuoss, R.M. Irreversible processes in electrolytes. Diffusion, conductance, and viscous flow in arbitrary mixtures of strong electrolytes. *J. Phys. Chem.* 1932, *36*, 2689-2778.

[137] Onsager, L. Z. The theory of electrolytes. *Z. Phys.* 1926, *27*, 388-392.

[138] Debye, P. and Hückel, H. Bemerkungen zu einem satze über die kataphoretische wanderungsgeschwindigkeit suspendierter teilchen. *Z. Phys.* 1924, *25*, 49-52.

[139] Jones, G.; Dole, M. The viscosity of aqueous solutions of strong electrolytes with special reference to barium chloride. *J. Am. Chem. Soc.* 1929, *51*, 2950-2964.

[140] Abdulagatov, I.M.; Azizov, N.D. Densities, Apparent Molar Volumes, and Viscosities of Concentrated Aqueous $NaNO_3$ Solutions at Temperatures from 298 to 607 K and at Pressures up to 30 MPa. *J. Sol. Chem.* 2005, *34*, 645-685.

[141] Abdulagatov, I.M.; Zeinalova, A.B.; Azizov, N.D., Viscosity of aqueous electrolyte solutions at high temperatures and high pressures. Viscosity B-coefficient. Sodium iodide. *J. Chem. Eng. Data* 2006, *51*,1645-1659.

[142] Abdulagatov, I.M.; Azizov, N.D.; Zeinalova, A.B. Density, apparent and partial molar volumes, and viscosity of aqueous Na_2CO_3 solutions at high temperatures and high pressures. *Z. Phys. Chem.* 2007, *221*, 1-38.

[143] Abdulagatov, I.M.; Azizov, N.D.; Zeinalova, A.B. Viscosities, densities, apparent and partial molar volumes of concentrated aqueous $MgSO_4$ solutions at high temperatures and high pressures. *J. Phys. Chem. Liquids* 2007, *45*, 127-148.

[144] Kaminsky, M. *Z.* Experimentelle untersuchungen über die konxentrations- und temperaturabhangigkeit der zähigkeit wäβriger lösungen starker electrolyte. I. Mitteilang: KJ-, NH4Cl- und Na_2SO_4-lösungen. *Phys. Chem.*1955, *5*, 154-191.

[145] Kaminsky, M. *Z.* Experimental study of the concentration and temperature dependence of the viscosity of aqueous solutions of strong electrolyte. I. *Phys. Chem.* 1956, *8*, 173-191.

[146] Kaminsky, M. *Z.* Concentration and temperature dependence of the viscosity of aqueous solutions of strong electrolytes. III. KCl, K_2SO_4, $MgCl_2$, $BeSO_4$, and $MgSO_4$ solutions. *Phys. Chem. N.F.* 1957, *12*, 206-231.

[147] Kaminsky, M. *Disc. Faraday Soc.* 1958, *24*, 171-179.

[148] Desnoyers, J.E.; Arel, M.; Leduc, P.A. Conductance and viscosity of n-alkylamine hydrobromides in water at 25 °C: influence of hydrophobic hydration. *Can. J. Chem.* 1969, *47*, 547-553.

[149] Desnoyers, J.E.; Perron, G. The viscosity of aqueous solutions of alkali and tetraalkylammonium halides at 25 °C. *J. Sol. Chem.* 1972, *1*, 199-212.

[150] Feakins, D.; Lawrence, D.G. The relative viscosities of solutions of sodium and potassium chlorides and bromides in N-methylformamide at 25, 35, and 45 °C. *J. Chem. Soc. A.* 1966, 212-219.

[151] Robertson, C.T.; Tyrrell, H.J.U. Viscosity B coefficients for some substituted anilinium cations in aqueous solutions at 25 and 35 °C. *J. Chem. Soc. A.* 1969, 1938-1942.

[152] Abdulagatov, I. M.; Zeinalova, A.B.; Azizov, N.D. Experimental viscosity B-coefficients for aqueous LiCl solutions. *J. Mol. Liquids* 2006, *126*, 75-88.

[153] Thomas, D.G. Transport characteristics of suspension: VIII. A note on the viscosity of Newtonian suspensions of uniform spherical particles. *J. Colloid Sci.* 1965, *20*, 267-277.

[154] Einstein, A. Berichtigung zu meiner Arbeit: "Eine neue Bestimmung der Molekül-dimensionen." *Ann. Phys.* 1911, *34*, 591-592.

[155] Breslau, B.R.; Millero, I.F. On the viscosity of concentrated aqueous electrolyte solutions. *J. Phys. Chem.* 1970, *74*, 1056-1061.

[156] Moulik, S.P.; Rakshit, A.K. A.K. Viscosity of aqueous electrolyte solutions as a function of *B*-coefficients. *J. Ind. Chem. Soc.* 1975, *52*, 450-453.

[157] Sahu, B.; Behera, B. Viscosity of concentrated aqueous solutions of I:I electrolytes. *Ind. J. Chem. A* 1980, *19*, 1153-1157.

[158] Altan, A.; Kus, S.; Kaya, A. Rheological behavior and time dependent characterization of gilaboru juice (*viburnum opulus L.*). *Food Sci. Tech. Int.* 2005, *11*, 129-137.

[159] Gochiyaev, B.R. Determination of the viscosity and density of apple juice. *Izv. Vyssh. Ucheb. Zaved., ser. Pish. Tekhnol.* 1964, *4*, 93-95.

[160] Saravacos, G.D. Tube viscometry of fruit purees and juices. *J. Food Science* 1968, *22*, 1585-1588.

[161] Schwartz, M.; Costell, E. Influencia de la temperature y la concentracion en la viscpsidad de los zumos de manzana y de uva. *Revista de Agroquimica y Technologia de Alimentos* 1989, *29*, 239-245.

[162] Moresi, M.; Spinosi, M. Engineering factors in the production of concentrated fruit juices. 1. Fluid physical properties of orange juices. *J. Food. Technol.* 1980, *15*, 265-276.

[163] Steverson, E.M. In: *Quality summary. Citrus juice products from FMC citrus juice extractors*. Citrus Machinery Div., FMC Corp.: Lakeland, Fla., 1977; 1978; 1979.

[164] Mizrahi, S.; Berk, Z. Flow behavior of concentrated orange juice: Mathematical treatment. *J. Texture Studies* 1972, *3*, 69-79.

[165] Mizrahi, S.; Berk, Z. Flow behavior of concentrated orange juice. *J. Texture Studies* 1970, *1*, 342-355.

[166] Mizrahi, S.; Firstenberg, R. Effect of orange juice composition on flow behavior of six-fold concentrate. *J. Texture Studies* 1975, *6*, 523-532.

[167] Ezell, G.H. Viscosity of concentrated orange and grapefruit juices. *Food Technology* 1959, *13*, 9-13.

[168] Varshney, N.N.; Kumbhar, B.K. Effect of concentration and temperature on rheological properties of pineapple and orange juices. *J. Food. Sci. Tech. India* 1978, *15*, 53-55.

[169] Hernandez, E.; Chen, C.S.; Johnson, J.; Carter, R.D. Viscosity changes in orange juices. *J. Food Eng.* 1995, *25*, 387-396.

[170] Dubois, C.W.; Murdock, D.I. The effect of concentration on quality of frozen orange juice with particular refernce to 58.5° and 42 °Brix products. Chemical and Physiological aspects. *Food Technol.* 1955, *9*, 60-63.

[171] Berk, Z. Viscosity of orange juice concentrates: Effect of ultrasonic treatment and concentration. *Food Technol.* 1964, *18*, 153-154.

[172] Rouse, A.H.; Albrigo, L.G.; Huggart, R.L.; Moore, E.L. Viscometry affected by pectic constituents in 45 °Brix frozen concentrated orange juice. *Ann. Meeting of the Florida State Horticultyral Society* 1973, *86*, 249-254.

[173] Ingram, M. *The viscosity of concentrated orange juice*; The Hebrew University: Jerusalem, 1948.

[174] Moresi, M.; Spinosi, M. Engineering factors in the production of concentrated fruit juices. II. Fluid physical properties of grape juices. *J. Food Tech.* 1984, *19*, 519-533.

[175] Khalil, K.E.; Ramakrishna, P.; Nanjundaswamy, A.M.; Patwardhan, M.V. Rheological behavior of clarified banana juice: Effect of temperature and concentration. *J. Food Eng.* 1989, *10*, 231-240.

[176] Fito, P.J.; Clemente, G.; Sanz, F.J. Rheological behavior of tomato concentrate (hot break and cold break). *J. Food Eng.* 1983, *2*, 51-62.

[177] Foda, Y.H.; McCollum, J.P. Viscosity as affected by various constituents of tomato juice. *J. Food Sci.* 1970, *35*, 333-338.

[178] Saravacos, G.D.; Oda, Y.; Moyer, J.C. *Tube viscometry of tomato concentrates*; NY State Agr. Exp. Station Bull., Cornell University: Geneva, NY, 1967.

[179] Harper, J.C.; El-Sahrigi, A.F. Viscometric behavior of tomato concentrates. *J. Food Sci.* 1965, *30*, 470-476.

[180] Molwane, S.D.; Gunjal, B.B. Viscometric characteristics of cold-break and hot-break extracted tomato juice concentrates. *J. Food Sci. Tech.* 1985, *22*, 353-357.

[181] Krapivin, N.I.; Fedotkin, I.M. *Konserv. Ovoshch. Prom.* 1977, *39*, 40-46.

[182] Luh, B.S.; Villarreal, F.; Leonard, S.J.; Yamaguchi, M. Effect of ripeness level on consistency of canned tomato juice. *Food Technology* 1960, *12*, 635-639.

[183] Giner, J.; Ibarz, A.; Garza, S.; Xhian-Quan, S. Rheology of clarified cherry juices. *J. Food Eng.* 1996, *30*, 147-154.

[184] Altan, A.; Maskan, M. Rheological behavior of pomegranate (*Punica granatum L.*) juice and concentrate. J. Texture Studies 2005, *36*, 68-77.

[185] Peleg, H.; Noble, A.C. Effect of viscosity, temperature and pH on astringency in cranberry juice. *Food Quality and Preference* 1999, *10*, 343-347.

[186] Vitali, A.A.; Roig, S.M.; Rao, M.A. Viscosity behavior of concentrated passion fruit juice. *Confructa* 1974, *19*, 201-106.

[187] Garciá, R.; Rivera, J.; Rolz, C. Rheological properties of some tropical fruit products and their enzymic clarification. *Proc. IV Int. Cong. Food Sci. Tech.* 1974, *2*, 18-26.

[188] Lopez, A.; Ibarz, A.; Pagan, J.; Vilavella, M. Rheology of wine musts during fermentation. *J. Food Eng.* 1989, *10*, 155-161.

[189] Oomah, B.D.; Sery, G.; Godfrey, D.V.; Beveridge, T.H.J. Rheology of sea buckthorn (*hippophae rhamnoides* L.) juice. *J. Agric. Chem.* 1999, *47*, 3546-3550.

[190] Lau, A.K.; March, A.C.; Lo, K.V.; Cumming, D.B. Physical properties of celery juice. *Can Agric Eng.* 1992, *34*,105-110.

[191] Manohar, B.; Ramakrishna, P.; Udayasankar, K. Some physical properties of tamarind (*Tamarindus indica L.*) juice concentrates. *J. Food Eng.* 1991, *13*, 241-258.

[192] Maskan, M. Rheological behavior of liquorice (*Glycyrrhiza glabra*) extract. *J. Food Eng.* 1999, *39*, 389-393.

[193] Gunjal, B.B.; Waghmare, N.J. Flow characteristics of pulp, juice and nectar of 'baneshan' and 'neelum' mangoes. *J. Food. Sci. Tech. India* 1987, *24*, 20-23.

[194] Sogi, D.S. Effect of concentration and temperature on the viscosity of watermelon juice. *J. Food Science and Technology* 2003, *40*, 509-511.

[195] Simpson, M. Viscosity as related to pectin content in strawberry juices. *Horticultural Experimental Station and Products Laboratory*; 1958, pp 126-129.

[196] Vitali, A.A.; Rao, M.A. Flow behavior of guava puree as a function of temperature and concentration. *J. Texture Srudies* 1982, *13*, 275-289.

[197] Brekke, J.E.; Myers, A.L. Viscometric behavior of guava purees and concentrates. *J. Food Science* 1978, *43*, 272-273.

[198] Manohar, B.; Ramakrishna, P.; Ramteke, R.S. Effect of pectin content on flow properties of mango pulp concentrates. *J. Texture Studies* 1990, *21*, 179-190.

[199] Singh, N.I.; Eipeson, W.E. Rheological behavior of clarified mango juice concentrates. *J. Texture Studies* 2000, *31*, 287-295.

[200] Watson, E.L. Rheological behavior of apricot purees and concentrates. *Can. Agr. Eng.* 1968, *10*, 8-11.

[201] Harper, J.C.; Lebermann, K.W. Rheological behavior of pear purees. *Proc. 1st Int. Cong. Food Sci. Tech.* 1962, *1*, 719-728.

[202] Elfak, A.M.; Pass, G.; Phillips, G.O.; Morley, R.G. The viscosity of dilute solutions of guar gum and locust bean gum with and without added sugars. *J. Sci. Food Agric.* 1978, *28*, 895-899.

[203] Soesanto, T.; Williams, M. Volumetric interpretation of viscosity for concentrated and dilute sugar solutions. *J. Phys. Chem.* 1981, *85*, 3338-3341.

[204] Rao, M.A. Rheology of liquid foods-A review. *J. Texture Studies* 1977, *8*, 135-168.

[205] Mizrahi, S. A review of the physicochemical approach to the analysis of the structural viscosity of fluid fruit product. *J. Texture Studies* 1979, *10*, 67-82.

[206] Jones, G.; Talley, S.K. The viscosity of aqueous solutions as a function of the concentration. II. Potassium bromide and potassium chloride. *J. Am. Chem. Soc.* 1933, *55*, 4124-4125.

[207] Charm, S.E. In: *The Fundamentals of Food Engineering*; Conn., Avi Pub., Co.: Westport, 1963; Chapter 3, pp 47-106.

[208] Bálint, A. Prediction of physical properties of foods for unit operations. *Periodica Polytechnica Ser. Chem. Eng.* 2001, *45*, 35-40.

[209] Bayindirli, L.; ve Ersan, C. Mathematical modeling of variation of density and viscosity of clear fruit juices with temperature and concentration. *Gida Teknolojisi* 1998, *3*, 66-69.

[210] Fernéndez-Martin, F. Influence of temperature and concentration on some physical properties of milk and milk concentrates. II. Viscosity. *J. Dairy Res.* 1972, *39*, 75-82.

[211] Peacock, S. Prediction physical properties of factory juices and syrups. *Int. Sugar Jnl.* 1995, *97*, 571-577.

[212] Elfak, A.M.; Pass, G.; Phillips, G.O. The viscosity of dilute solutions of ϑ-carrageenan and sodium carboxymethylcellulose. *J. Sci. Food Agric.* 1978, *29*, 557-562.

[213] Odigboh, E.U.; Mohsenin, N.N. Effect of concentration on the viscosity profile of cassava starch pastes during the cooking-cooling process. *J. Texture Studies* 1975, *5*, 441-457.

[214] Geller, V.Z.; Pugach, A.K.; Mkhitaryan, G.G. Density and effective viscosity of apple juice at various concentrations. *Pishevaya Promyshlennost* 1992, *1*, 45-56.

[215] Holdsworth, S.D. Applicability of rheological models to the interpretation of flow and processing behavior of fluid food products. *J. Texture Studies* 1971, *2*, 393-418.

[216] Autio, K.; Houska, M. Measurement of flow curves for model liquids and real food systems with two commercial viscometers. *J. Food Eng.* 1991, *13*, 57-66.

[217] Alviar, M.S.B.; Reid, D.S. Determination of reological behavior of tomato concentrates using back extrusion. *J. Food Science* 1990, *55*, 554-555.

[218] McCarthy, K.L.; Seymour, J.D. A fundamental approach for the relationship between the bostwick measurements and Newtonian fluid viscosity. *J. Texture Sciences* 1994, *24*, 1-10.

[219] Erickson, E.R.; Berntsen, R.A.; Eliason, M.A. Viscosity of corn syrup. *J. Chem. Eng. Data* 1966, *11*, 485-488.

[220] Rai, P.; Majumdar, G.C.; DasGupta, S., De, S. Prediction of the viscosity of clarified fruit juices using artificial neural network: a combined effect of concentration and temperature. J. Food Engineering 2005, 68, 527-533.

INDEX

D

E

F

G